It's all here—true tales of
misadventure, tragedy, and courage from the annals
of U.S. history

Read about:

*The Coconut Grove Fire of 1942

*The Port Chicago Explosion of 1944

*The *Sultana* Explosion of 1865

*The Tri-State Tornado of 1925

*The Chicago School Fire of 1958

. . . and more

**ACE COLLINS** is an author and radio show host who has appeared on CNN, *CBS This Morning*, *The NBC Nightly News*, *Entertainment Tonight*, and *Good Morning America*. He lives in Texas.

# TRAGEDIES OF AMERICAN HISTORY

## 13 STORIES OF HUMAN ERROR AND NATURAL DISASTER

ACE COLLINS

A PLUME BOOK

PLUME
Published by the Penguin Group
Penguin Group (USA) Inc., 375 Hudson Street,
New York, New York 10014, U.S.A.
Penguin Books Ltd, 80 Strand,
London WC2R 0RL, England
Penguin Books Australia Ltd, 250 Camberwell Road,
Camberwell, Victoria 3124, Australia
Penguin Books Canada Ltd, 10 Alcorn Avenue,
Toronto, Ontario, Canada M4V 3B2
Penguin Books (N.Z.) Ltd, Cnr Rosedale and Airborne Roads,
Albany, Auckland 1310, New Zealand

Penguin Books Ltd, Registered Offices:
80 Strand, London WC2R 0RL, England

First published by Plume,
a member of Penguin Group (USA) Inc.

First Printing, June 2003
10 9 8 7 6 5 4 3 2 1



LIBRARY OF CONGRESS CATALOGING-IN-PUBLICATION DATA

Collins, Ace.
Tragedies of American history : 13 stories of human error and natural disaster / by Ace Collins.
p. cm.
ISBN 0-452-28300-0 (pbk.)
1. Disasters—United States—History—Anecdotes.
2. United States—History—Anecdotes. I. Title.

E179 .C65 2002 2001047395
363.34'0973—dc21

Printed in the United States of America
Set in Simoncini Garamond
Designed by Leonard Telesca

*The book is dedicated to the Red Cross*
*and Salvation Army, who are always there*
*when people have been caught in a disaster*
*and have no one to whom they can turn.*

# Contents

# Foreword

This book deals with ordinary people whose security was forever shattered after being touched by traumatic events. A few of these calamities could not have been avoided; they were truly acts of God. Yet most of the historical episodes covered in this book did not have to happen. Lives were lost due to carelessness, greed or ignorance. In retrospect, in nearly every case, be it an act of God or an act of man, almost anyone should have been able to realize that disaster was about to strike, that they were about to be tossed into harm's way. So why didn't the victims react? Why didn't they take the steps to avoid catastrophe?

Like a majority of Americans, most were blinded by thinking that they were safe, that nothing bad could possibly happen to them, that bad things only happened to other people. With these thoughts clouding their vision, they didn't look for the easy-to-read warning signs. They didn't evaluate each step they took in life.

The most horrible facet of these stories is that they occur in places where the victims felt completely safe. With the possible exception of the events involving a submarine trip or the fury of a hurricane, most of us would have felt safe too. We don't worry when we are locked in our homes, when we board a passenger jet, take a cruise, walk into a school, shop in a store or hear a storm warning. We feel secure in these normal situations. Thus, we don't seek out the exits in nightclubs and restaurants, don't listen to the instructions for surviving crash landings on planes or carefully follow storm-tracking radar.

The people in these stories acted in the same way. They never thought that they would find themselves in a moment-to-moment struggle for life. They never believed that they would be the ones whose security would be dashed in just seconds. They never speculated that they would have to find an exit in a fire or know how to get off a crashed plane or a sinking ship. Ultimately they seemed to have faith that even if something did go horribly wrong, that someone else would guide them through the trouble, someone else would know what to do, someone else would save them. They simply did not prepare for the most important moment in their lives, and sadly, few of us do either.

The situations addressed in this book include only a small portion of the monumental tragedies that have hit America. Those other stories will have to wait. We have chosen to touch on thirteen incredible moments when something went dramatically wrong and time seemed to stop. Thirteen different situations that read like fiction, but are really unchangeable fact. Thirteen different portraits of terror and fear that might mirror the works of Alfred Hitchcock or Stephen King, but deal with real people caught in very real events.

As you read these stories, try to put yourself into each horrific incident. Consider if you would have felt safe in the moments before disaster struck. Think about how you would have reacted when challenged by forces beyond your control. Question if you would have been prepared to react wisely and if you would have had what it took to survive.

None of us knows what the future holds. Most of us believe that we will never board the plane that crashes or be buried under tons of floodwater, but those things happened all the time. And, as history has shown, they have happened even to those who thought they were safe.

# TRAGEDIES OF AMERICAN HISTORY

# 1

# Senseless Tragedy over New York 1960

Late in the afternoon of December 15, 1960, famed New Zealand adventurer Sir Edmund Hillary, who had been the first to conquer Mt. Everest just seven years before, fielded yet another request for a public appearance. For several days he had been hailed as a hero by thousands in Chicago, and now, just when he thought he had finished this round of speeches and fund-raisers, another group called begging him to speak to them the next day. Though he was tired and desperately needed some time away from the press and public, Hillary checked his upcoming New York schedule and consulted with his two traveling companions, Sherpa Khumjo Chumbi and writer Desmond Doig. When he discovered he had nothing planned on the East Coast for several days and his friends offered no objections to staying in Chicago for another day, reluctant Hillary agreed to speak to one more Midwestern group.

After accepting the new engagement, the mountain climber called the United Airlines reservation desk and canceled the three seats he had booked on a Friday-morning New York flight. As fate would have it, this last-moment change in plans would not only save the touring trio's lives, but would allow an eighteen-year-old college student the opportunity to purchase a seat on the previously sold-out Flight 826.

While Sir Edmund slept, Peggy La Riviere, Alvin Sokolsky and Stephen Baltz woke up about the same time on December 16. Certainly they were all excited too. Peggy, a student at Barat College in Lake Forest, Illinois, a suburb of Chicago, wanted to get back home to Greenwich, Connecticut, for the Christmas break. Her father, Milton, had come to visit her during the week and the two had planned to return east together. Yet numerous calls to the airlines had secured the La Rivieres just a single seat. It was only when Sir Edmund Hillary canceled that United Airlines had called Peggy and gave her the seat she desperately needed. Though La Riviere would be leaving O'Hare Airport a few minutes behind her father, she was happy knowing that they would be quickly reunited at the New York International Airport. Then it would be just a short drive home to the rest of the family she had not seen since September.

Unlike La Riviere, frequent business traveler Alvin Sokolsky had made his reservations for Flight 826 early. Alvin, who had been married only a few weeks, was anxious to get back to his bride in Baltimore. He knew that his Ruth would be waiting at the terminal in New York. The sooner he could be in her arms, the better he would feel. Within moments of when Peggy La Riviere was dashing from her dorm room, Alvin Sokolsky, a determined glint in his eye, was saying good-bye to his hotel room and hailing a cab.

Like the college student and the businessman, eleven-year-old Stephen Baltz of Wilmette, Illinois, had also bounced out of his bed before dawn that Friday morning. His mind was so filled with images of Christmas in New York he had scarcely been able to sleep the night before. After carefully putting on a new gray suit and checking and rechecking the suitcase that had now been packed for almost a week, Stevie, as his parents called him, took out the plane ticket and studied it carefully.

Like La Riviere and Sokolsky, Baltz was booked on United Airlines Flight 826. Leaving Chicago just after 8:00 A.M., the shiny DC-8 jet was scheduled to land in New York City by eleven Eastern time. For a boy headed to his grandparents' home, flying on a jet was a Christmas wish come true.

Stevie's father, William Baltz, vice president of the Admiral electronics corporation, watched with great pride as his redheaded son put on a new gray hat, adjusted the single feather that had been stuck in the hat's band, then gleefully stared at the image reflected in a mirror. It was hard for William to believe that just two days before this same face had been covered with tears. A case of the flu had kept Stevie from taking the holiday trip to New York with his mother, Phyllis, four-year-old brother, Willie, and ten-year-old sister, Randee. They had been forced to fly to New York without him. Now, fully recovered, the boy couldn't wait to catch up with his siblings. As fate would have it, his illness would cost him a great deal more than just two days in New York.

As William and Stevie drove through the Friday-morning rush-hour traffic, the boy excitedly spoke about Christmas. Once they had parked at the airport and entered the terminal, Stevie could barely contain himself. Carols played over the public address system, garlands and lights hung everywhere, a jolly mechanical Saint Nick waved from one corner and even the ticket agents joined in the spirit of the season, merrily telling everyone to have a happy holiday. To the boy, the day was going to be the best day of his life.

Stevie and William waded through a mass of travelers before finally arriving at the departure gate for Flight 826. When they walked out into the crisp air, the boy spotted the most beautiful airplane he had ever seen. White and silver, sporting four massive jet engines, the sleek DC-8 Super 62 looked as if it could literally leap into the air on a moment's notice. It made the turbo-prop planes sitting on either side of it seem as out-of-date as Model-A Ford sedans. As Stevie's eyes feasted on the mass of modern machinery, he tried to imagine the space-age technology that must have been placed in the cockpit. He was hoping that he could somehow sneak a peek into the pilot's domain once he entered the jet.

For William, observing his son marveling at the jet almost made up for the boy having to wait at home an extra two days. The father couldn't imagine a better Christmas present for Stevie than getting to take a solo ride on a DC-8. The only way this trip could be better would be if he was able to fly by his son's side.

The two said good-bye at the departure gate and William watched the boy climb the steps to the jet's cabin. Not far behind Stevie were Peggy La Riviere and Alvin Sokolsky. While Peggy and Alvin quickly looked for their seats, Stevie stopped and peered through the partially opened cockpit door. He was amazed by the control panel. He had never seen so many dials or levers. He wondered how any single person could know what all of the readouts meant. He also wondered if all of them were really necessary. If one wasn't working, would it really matter?

Besides Stephen Baltz, Peggy La Riviere and Alvin Sokolsky, seventy-four other passengers boarded the United Airlines DC-8 that cold winter morning. Somehow Mrs. Edwige Dumalshi, of Oak Lawn, Illinois, had gotten her son and daughter, Patrick and Joele, ready and on the flight. Carlos and Olga Wittmer, of Caracas, Venezuela, had bundled up their infant son and made the flight as well. So had Susan Gordon, Beverly Parks, and Earl Reems. As one by one the seventy-seven ticketed passengers got comfortable and waited for the DC-8 to take off and turn toward the east, William Baltz left the O'Hare Airport terminal and drove to his office, confident in the knowledge that his wife would soon be picking Stevie up at New York International.

As Flight 826 left the frozen Illinois ground behind and the flight attendants went over the emergency procedures, Stevie fought back his fears of flying alone and thought of the trip ahead of him. Within hours he would be seeing the huge Christmas tree at Rockefeller Center, Santa Claus greeting children at Gimbles, families strolling through the snow in Central Park, and thousands of holiday shoppers flocking to stores made famous by books and motion pictures. This promised to be the Christmas adventure of a lifetime for the boy, one wrapped in more excitement and surprises than his young mind could imagine. As the fields of Indiana, framed by the window beside his seat, spread out as far as he could see, Stevie impatiently counted the minutes until his mother met him at New York International. Checking the watch he proudly wore on his left wrist, he noted the time as fifteen past eight.

Not more than a hundred feet in front of the boy, a calm but boisterous United flight crew made up of forty-six-year-old Captain R. J. Sawyer, First Officer R. W. Piecing, forty, and the baby of the crew, thirty-year-old Flight Engineer R. E. Prewitt, told jokes, drank coffee and talked about their own holiday plans. All based out of California, they were a confident crew. Their powerful plane was performing at peak efficiency, and because they were riding a strong jet-stream, Flight 826 looked ready to hit the Big Apple a few minutes ahead of schedule. In the minds of the crew this was the only way to travel. The DC-8 was so advanced, many pilots boasted, that the plane didn't even need a pilot. So for Sawyer, Piecing and Prewitt, it looked like another easy day at the office.

As the pilot easily guided the big bird across the snow-blanketed Midwest, the crew expressed little concern that one of the plane's two VOR navigation receivers, the instruments that helped determine position, was not functioning. Being without half the emergency system necessary for instrument landings in congested traffic didn't even cause the men to pause and consider the weather conditions that lay ahead. Their standing joke was that United would fix the VOR as a Christmas present for the crew, though the seasoned team thought they would probably never need it anyway.

Except for attendants Ann Marie Bouthan, Augustine Ferrer, Mary Mahoney and Pat Keller constantly checking on Stevie, offering him cokes and peanuts, as well as asking the boy about his ever-growing Christmas wish list, to the passengers on board 826, the flight seemed routine. In the plane's crowded cabin, a few passengers visited with each other, some read magazines, others slept, and the remainder stared outside into conditions that seemed more like Christmas with each mile the aircraft drew closer to New York.

About the same time Stevie Baltz was sipping on his first soft drink, a few hundred miles away a big four-engine prop-driven Super Constellation was touching down in Columbus, Ohio. The TWA Flight 266 crew had picked up their plane in Cleveland, then made a stop for additional passengers in Dayton and now in Columbus. The "Connie," as it was called by those in the aircraft industry, was one of the most recognizable ships in the air. Its unique three tail fins gave the plane a wild "triple wing" look. Yet compared to the DC-8 jet, the Constellation, sporting half the instruments and its propeller blades, appeared old-fashioned. As aircraft became more modern, this seemed to be a plane of the past, not the future.

At Dayton, only three passengers had come across the tarmac to the TWA stairs: Cecil Mullins, his wife, and their infant. The family was on their way to the East Coast to introduce the youngest Mullins to her grandparents. The parents were full of excitement at going home for their baby's first Christmas.

In Columbus, a trio of Ohio State students, Robert McEschern, Richard Magnuson and Maurice Bricker, eagerly charged up the loading steps first. They had finished school for the year and were on their way to New York for the holidays. Following the trio of close friends were a number of other students and businessmen. Many of the men in suits, like A. B. Swenson, were employed by the aircraft industry. They flew all the time and had no concerns about this flight being any different from the scores of others they had experienced.

In the Connie's cockpit, the New York–based team of Captain David Wollan, First Officer Dean Bowen and Flight Engineer Leroy Rosenthal rechecked their instruments then pointed the big bird toward LaGuardia. Though the tower had warned them about light snow and drizzle at their destination, none of the three was worried. They had complete confidence in their ability to guide the plane in safely for an instrument landing.

Back in the plane's main cabin, Patricia Post and Margaret Gernat began to serve the passengers soft drinks and coffee. Though only twenty-one and twenty-four, the flight attendants easily addressed each of their duties like seasoned professionals. Asking as many questions about Christmas expectations as they answered inquiries about matters concerning the flight, the two women quickly got to know the thirty-nine passengers who were flying with them. Neither the flight attendants nor most of the passengers noted the thick line of clouds as the Connie crossed the Pennsylvania state line.

Several hundred miles away in New York, the city was working through its first big winter storm of the year. With twinkling lights and colorful decorations carefully placed in shop windows, the city embraced the holiday season with a youthful excitement and optimism fueled by the recent election of John F. Kennedy to the White House. To young and old alike, this was a time to look to the future with the hope that, like the jet age, America's best days were just arriving.

Not far from the city's finest shops and legendary stores, in one of the oldest sections of Brooklyn along a once proud row of brownstones and shops, Joseph Colacano and his uncle, John Operisano, had just opened a small sidewalk Christmas tree lot. Peppered with a mix of snow, sleet and drizzle, the men laughed and wished passersby a Merry Christmas. Though Sterling Place, where Joe and John had set up their seasonal business, was now a struggling, almost impoverished area, the snow that had begun four days before stirred the holiday spirit within the little run-down neighborhood. Everyone who greeted the men that day appeared filled with holiday cheer.

With shovel in hand, Charles Cooper waved at Colacano and Operisano as he walked by their Christmas tree display. At thirty-four, Cooper was employed as a city sanitation worker. Instead of dealing with trash this Friday morning, he had been assigned to clear snow from the Brooklyn sidewalks and streets. Though it was hard work, Cooper relished the opportunity for the change of pace and a chance to visit with those who passed by as he worked. Scooping snow was much better than hauling trash.

Cooper had been shoveling for more than an hour when Henry McCaddin got up. McCaddin was happy to see the man with the shovel. Henry, his wife and their one-year-old son not only lived just above where Cooper was working, but also operated the McCaddin Funeral Parlor, located a floor beneath their tiny apartment. McCaddin knew the time just before Christmas was very difficult for mourners, and when the streets were clogged by snow, the grieving seemed to take an even greater toll on those who came to his establishment. Getting rid of the huge drifts would at least solve one of his problems.

Across from the funeral home Wallace E. Lewis toiled alone. At ninety, he was still having to work to make ends meet. As the caretaker of the Pillar of Fire Methodist Church, he lived in a room in the old house of worship and did all the cleaning, yard work and general repairs for the building. His age and the seeming enormity of taking care of a three-story structure hadn't dimmed his zest for life or his generous spirit. Lewis would throw open the gates to the churchyard each day so local children would have a safer place than the streets to play. Even as the snow and drizzle hit his window, the old man was anticipating another group of youngsters knocking on his door as soon as school was out. He was going to have to begin early if he was to have the snow scooped from the playground equipment by three.

Just down the street from the old stucco church building, James Barnes, a junior high teacher at Public School 614, was trying to get his class focused on their lessons. It seemed a futile effort. The twenty-five students who had crowded into his room were passing notes and looking out the window at the snow falling on Sterling Place. For a moment, the frustrated teacher considered closing the blinds, but with winter vacation just around the corner, he figured even that wouldn't help him hold his class's attention today.

Constance Clazzo was spending the morning at her flower shop attempting to decorate her front window. Using flowers she had in stock as well as old Christmas decorations, Mrs. Clazzo was so intent on the task at hand, she barely noticed the falling snow had now become mixed with a light freezing drizzle.

Just a few feet from Clazzo's front door, Dennis Doery, Dimian Reyes and Jimmy Triczono were all involved in remodeling projects.

Though working on separate jobs, the three men knew each other by sight. Each was also hoping that their small jobs would grow into a massive rebuilding of the run-down neighborhood.

Eight miles away, just across the Narrows of New York Harbor, in a neat middle-class housing subdivision, a Staten Island housewife, Maude Petersen, was trying to get her husband off to work and her children dressed for school. She had a million things she wanted to accomplish and she knew she wouldn't get started on her list until she had her family out from under foot.

Less than a mile away from the Petersens' house, servicemen had been working since dawn clearing the snow from Miller Army Air Field. To many the small base seemed outdated in the jet age, yet lying in the heart of New York, Miller remained a strategic point for the armed forces. Still, on December 16, the soldiers didn't expect anything more exciting than the arrival of a few small prop planes.

Peter and Gerard Paul were within a stone's throw of Miller Field shopping at a newly opened strip center. With Christmas just around the corner and a long list of gifts to buy, the brothers had mapped out this day to purchase everything they needed for the season.

Though he didn't know the brothers, Clifford Beauth, a driver for an oil company, met them on a Staten Island street as he slipped and slid along his delivery route. With the onset of cold weather, fuel demands were high. For Beauth this promised to be a long day.

As the city settled in for what should have been a normal Friday morning, Flight 826 approached New York and received final approach instructions from the New York International control tower.

"United eight-two-six, now descend to and maintain five thousand."

"Maintain five thousand," the pilot answered. "We're leaving fourteen. United eight-two-six."

"Roger," a controller replied as he studied his radar. "Looks like you'll be able to make the Preston at five."

"Er," came the reply, "will head it right on down—we'll dump it."

"United eight-two-six, if holding is necessary at Preston, use Southwest one-minute pattern."

"Roger. We're out of seven."

"Eight-two-six," the controller shot back, "and you received the holding instructions at Preston, radar service is terminated, contact Idlewild [New York International] approach control one two three point seven. Good day."

"Good day."

As they set the large jet to flying a circular holding pattern, the crew casually checked the instruments and waited their turn to land. Even as they were locked in low clouds and unable to see the city less than a mile below, the three men were still relaxed and jovial. This leg of their journey was almost over. After a night of rest, they would be guiding a west-bound jet toward home.

As the pilots put the DC-8 into the holding pattern, Stevie Baltz peered out the window trying to pierce the fog and drizzle and catch a glimpse of the Empire State Building or the George Washington Bridge. Yet all he saw were brief flashes of the city. The weather had literally socked the town in. It was the first time today that the boy had been visibly disappointed.

Twelve miles away, the Super Constellation was in touch with the LaGuardia tower. Like Sawyer, Captain Wollan had been given the order to circle in a predefined holding pattern above the city. After leveling his turbo-prop plane at five thousand feet, the Connie's pilot rechecked his equipment and instruments. With visibility so poor, he fully expected to have to make a "blind" landing.

As the flight attendants on both planes gathered up cups and glasses and made sure the passengers were buckled in and ready to land, scores of family members and friends were gathering at the arrival gates for the two planes. In the International terminal, where Stevie Baltz's mother and sister had just arrived, it seemed like Christmas morning. Hurrying in at almost the same moment as the Baltzes were Milton La Riviere and Ruth Sokolsky. Holiday music was playing, people were smiling and decorations were everywhere. Those waiting for what was supposed to be only a few moments at a terminal gate were about to share an experience none of them would ever forget. It would lock them together for as long as they lived. Yet now, as they waited for the 826 to land, no one expected that they would ever see anyone in this waiting room again.

On the DC-8, Stevie Baltz checked his watch. It read 9:28. He hadn't reset it yet. It was still on Central Time. As he glanced back out the window, the clouds and mist suddenly cleared. In a breathtaking scene that reminded the boy of a postcard, he finally saw New York wearing all its Christmas finery. As one of the attendants passed him, Stevie exclaimed, "It looks like a fairy tale!"

Five thousand feet below, paying no attention to the skies, shoppers walked across snow-covered sidewalks past Santa Clauses ringing bells, store windows decorated with garlands and twinkling lights, as well as street vendors selling everything from hot chocolate to Christmas trees.

At the LaGuardia tower, a traffic controller followed the Connie through its holding pattern. He expected to see nothing that he hadn't noted a thousand times before. Yet as he watched the radar track along the screen, he suddenly saw a second unidentified blip heading toward the first plane at a high rate of speed. There was nothing routine about that. Not waiting to confirm who or what the unidentified blip was, the controller hurriedly called the TWA crew.

"Target at three o'clock—six miles!"

Checking their own radar, one of the men coolly shot back, "Got it." Though the crew was brought to a state of preparedness, none of them expected any real problems.

At LaGuardia, the controller was not as calm. Now the blip wasn't just approaching the Connie, it seemed to be racing toward it. It has to be at a different altitude, he thought. As the two blips came closer, he stood up and reopened the channel to Captain Wollan.

"Three miles—two o'clock!" the controller solemnly announced, wishing his radar could measure altitude as well as direction. Then, waiting for a response, he silently stared at his screen as the two blips became one. In agony he froze for a few seconds as he watched for visual confirmation that the planes had crossed paths at different heights and then reappeared as two separate images. Yet as the seconds ticked by, only one bright light illuminated his screen. "No!" he whispered. "God no!"

In the TWA cockpit, Wollan, Bowen and Rosenthal tried to peer through the clouds that surrounded them. Unable to see anything, they glanced from their instruments to the skies again and again. Their instruments assured them they were not alone, that there was another plane close to them, but where? Then, suddenly, without warning, they saw the jet. Moving at more than three hundred miles an hour, the second plane appeared to be nothing more than a blur of metal to the TWA crew, but they knew what it was. There was no time to think, no time to speak, not even time to pray. Jerking the wheel, Captain Wollan pulled the Connie into a fifty-degree bank. It was a desperate move that sent his half-filled coffee cup across the cockpit, knocked the flight attendants off their feet and shook numerous pieces of luggage from the overhead bins in the passenger compartment, but it was the only chance the plane had.

On the DC-8, no one was aware of anything in their path. The crew didn't know that their plane had somehow wandered eleven miles off course and was flying more than one hundred miles an hour faster than was designated for their holding pattern. With a VOR out of commission and roaring blindly through thick banks of clouds, the pilot only realized his flight was in trouble when he felt his plane smash through the top of the Connie. By then it was much too late.

As the United crew quickly tried to ascertain what had just happened, the Connie's pilot valiantly fought to control his plane. The Super Constellation's hull had been ripped wide open by the jet's wing. A jet engine, still rotating at thousands of RPMs, had fallen off the DC-8 and landed inside the passenger section of the prop plane. In front of scores of horrified people, a woman had been sucked into the engine's blades and chopped to pieces. The jet's roar drowned out the cries and screams of almost forty people as they watched the passenger disappear in front of them.

In the air above Staten Island, the deeply wounded TWA Connie was fighting to stay aloft. As Captain David Wollan heroically fought the all-but-useless controls, the ship sputtered, then began to streak toward the ground. Miller Field was directly below. The pilot still believed that if he could just regain some kind of control, he might make an emergency landing there.

Staring into walls of clouds, barely able to see the ground, the TWA crew attempted to will the plane across the New York landscape.

"Come on, baby," Wollan pleaded as the Connie slipped closer to the ground. Then, as if struck by a missile, the old ship shuttered. An engine blew and a wing caught fire. Sweat poured down the captain's face and the screams of passengers filled his ears, then another engine let go.

"God no," Wollan muttered, but refused to give up. Miller Field was now so close he could see it through the snow and haze.

As the wounded plane pushed toward the only safe haven available, Maude Petersen, just finishing the last of the dishes, heard the second engine explode. Looking from the window over the sink, she first saw smoke, then in the midst of the smoke she made out the outline of a huge plane struggling to stay airborne. It seemed so close to her home she thought it was going to hit her. Frozen, Petersen helplessly stared at the Connie as another explosion ripped the right engine from the plane's wing. It seemed the huge airship was falling apart a piece at a time.

As his plane fell apart around him, literally cracking and tearing down the middle of the fuselage, Wollan knew it was over. Unable to save his own life or the lives of his passengers, the thirty-eight-year-old pilot vowed not to come down on top of any houses or a school that lay right under him. His final few moments of life were spent trying to aim the helpless airliner at an open field.

While passing over the Petersens' house, the Connie blew its last engine. Hanging a thousand feet above the city, the plane erupted into a huge fireball, tossing much of its human baggage into the air. Many ejected from the TWA flight were literally blown to bits, but others, some still strapped in their seats, were alive and aware. For them, those last few seconds of life as they watched the snow-covered city rush up to meet them would bring terror like no one could imagine.

Peter and Gerard Paul heard the final explosion as they walked out of a store at the shopping center. In shock they stared as the huge plane broke into pieces and crashed to the ground. All that separated the brothers from the smoldering carnage was a chain-link fence and a hundred yards of snow-covered grass. Tossing their sacks of Christmas presents into the backseat of their car, the brothers raced to the fence, scaled it and ran toward the flaming, twisted hulk that had once been an airplane.

When Clifford Beauth heard the plane's final gasp, he pulled off to the side of the road and watched through his truck's windshield as people and wreckage fell into a snow-covered field just ahead of him. Feeling the pull to help, he headed his truck in the direction of Miller Field. As he got closer to where he judged the main section of the plane must have landed, he saw the body of a young man hanging in a tree. Getting out of his vehicle, Beauth was suddenly aware that pieces of plane, luggage and body parts had fallen all around him. In the street, just ahead of his truck and surrounded by insulation and clothing, was a wrapped Christmas package, its ribbon still neatly tied. Sick to his stomach the man turned away, leaned over the hood and cried for a moment. Then he got back into the truck's driver's seat and rushed to the field to join the Paul brothers as they searched for survivors.

As the three civilians peered through the flames, airmen at Miller Field were alerted that something was tragically wrong. Rushing outside, they smelled smoke and fuel. Loading into any vehicles that were available, they rushed by groups to the wreckage and began pouring water on the fire. Meanwhile the tower notified local hospitals, police and rescue teams that there appeared to have been a major air crash in the middle of Staten Island. Within minutes hundreds would be picking through debris looking for anyone who might have somehow survived.

Back on the still airborne DC-8, Stevie Baltz was trying to get another look at New York when he felt the huge plane shudder and heard what sounded like an explosion. Glancing around the cabin, he caught the eye of attendant Pat Keller. He could tell by the look etched on her face that she was as confused as he was. He could also tell she was in a state of shock.

For a moment, as the attendant and the boy stared at each other, the jet seemed suspended in the air, then, like a roller coaster, the plane shuddered again and began to dive toward the ground. For a few seconds there were no sounds except the struggling engines and the rushing air. Then, as more and more of the passengers became aware something was very wrong, many began screaming, a few praying and some cursing. In the cockpit the crew fought the controls, pleading with the big jet to push on. Yet with an engine ripped away and fuel pouring from a shredded gas line, the men's efforts were based more on hope and a quest for self-preservation than logic. The earth and the city below were coming closer to winning this final tug-of-war with each passing moment. With a wing all but sheered off, the big bird was caught in a fruitless struggle to stay aloft.

Inside the plane, as the passengers' eyes began to water due to the loss of pressure and the smell of leaking fuel, men and women looked to the flight attendants for some kind of advice. Yet there was none coming. The trio of professionals knew it was too late to go over the crash procedures that had been given as the plane took off. Now there was only time to pray. Within a few seconds the jet limped eight miles from the point of impact then rolled over a final time and dropped from the sky.

Just a few hundred feet below, carpenter Dennis Doery heard an explosion. Shielding his eyes against the falling drizzle, he looked up from the back of the store where he was working. Out of the clouds pieces of metal, some forty feet long, began showering down like rain. Racing for cover, Dennis watched as a large chunk of twisted wreckage flattened a car, then as another huge piece struck the Christmas tree stand where a confused and unaware Joseph Co-

Iacano and John Operisano were working. Charles Cooper was the next bystander Doery saw struck by the falling debris. Before the construction worker could become the next victim, he rushed for the cover of a grocery store.

At the same moment Dennis Doery was running for his life, James Barnes was trying to get his students to study the text on the practical application of history. Again the teacher noted one of his class members looking out the window. Barnes started to single out the guilty party, but the boy's pale face forced the man to turn his head in the same direction as the student's. Just outside the classroom window huge shards of steel were pelting buildings and people. What is that? the teacher thought. Then he realized that intact human bodies and body parts were raining from the sky like a bloody blizzard. Sheets of fire were everywhere. My God, he thought, we are in the middle of a war!

As shoppers and workers raced out of buildings and onto the streets, the main section of the plane came down like a bomb. It impacted nose first, the cabin opening up like a tin can, tossing passengers from its broken hull into streets and against building walls. Plowing through an apartment house, the now dead jet sprayed fuel across walls and streets before crashing into the Pillar of Fire church. Wallace E. Lewis never had a chance: He was dead even before a rapidly spreading blaze quickly consumed the building he called home.

As the metal and fire fell all around her flower shop, Constance Clazzo watched in horror. Too scared to move, she stood frozen at the window as a teenager ran by her, his clothes on fire. It seemed as though the world itself was ending. Brooklyn suddenly looked like a battle zone.

Even before anyone knew the cause, the fire and destruction set off city-wide fire and police alarms. Under the guidelines set forth in New York's civil defense plan, hundreds of units responded. Along with the firemen and police, teams of doctors and nurses rushed to Sterling Place. Even the Red Cross received an immediate call for relief aid. Yet with the piles of snow and traffic jams the wreckage caused, few firemen and policemen and almost no doctors could immediately get to the scene. In the moments after impact, it was up to the people of Sterling Place to not only save themselves, but to save anyone who might have survived the fiery disaster.

Dimian Reyes fought through walls of fallen bricks and metal wreckage to get to the largest remaining piece of the jet. As he looked into the hull, he saw a young woman stand up and peer out from the plane's cabin. Waving at her, Reyes began to run across the street to help her. Then, just when he was approaching the wreckage, she raised her arms over her face. Stopping in his tracks only a few yards from the section of the DC-8, the laborer watched in horror as an explosion tore the woman to pieces.

Jimmy Triczono pushed by the shocked Reyes and forced his way through snowbanks and debris to get to another large section of the plane. There, next to a long row of brick apartment houses, the man stared in amazement as a small boy, dressed in a smoldering suit, stumbled out of a broken section of the cabin. As he fell to the ground, his body suddenly grew lifeless. Yet as Triczono moved closer, the child cried out.

Policemen and firemen quickly joined Triczono. Stevie Baltz, still not completely aware of where he was or what had happened, lifted his head and asked, "Am I going to die?" His coat jacket still smoldering, his body twisted and burned, the young boy's question went unanswered. The child did appear more dead than alive. Yet in a sea of bodies, he was the only one they had seen who was breathing. As a light freezing rain began to fall and the men tried to ease his pain, a crowd of people gathered around the boy, not only to shield him from the elements, but to pray. Stevie, in obvious pain, whispered, "Mommy is waiting for me."

In his funeral home, Henry McCaddin observed the spreading flames and realized that not only was the Pillar of Fire church completely engulfed, but his business and the rest of the block was in danger as well. Gathering his family, McCaddin rushed out of his building and into the streets. Telling his wife to find safety at a friend's, the man so used to death waded through the refuse in a vain attempt to find life. "My God," he sighed, "everyone is dead!"

At Miller Field the Connie's crew had stayed with the plane long enough to keep her from crashing into a school or a subdivision, but the men could do nothing to keep the TWA flight's passengers out of harm's way. In trees and fields, in backyards and on city streets, the bodies had landed on Staten Island. Though four people, a woman and three men, would be breathing when volunteers and army personnel found them, only one would live long enough to die at the door of a hospital. Of the 135 passengers who flew on both planes that day, only Stevie Baltz had made it to New York alive.

As the time for Flight 826's arrival passed, Randee Baltz asked her mother why it was taking so long. A worried Phyllis looked back at the gate attendants as they whispered among themselves. Hovering beside the Baltzes, Ruth Sokolsky and Milton La Riviere also began to sense something was wrong. Along with scores of others they stared at the empty gate. Every few moments a uniformed ticket agent would take down the arrival time and place a later one on the board. Finally, after an agonizing thirty minutes, the arrival time was removed and nothing was added in its place. An airline official then asked those hovering around the gate to follow him into an empty meeting room. A few moments later men and women streamed sobbing from that same room. They had finally learned the truth.

A helpless Phyllis Baltz took her daughter's hand and led her through a crowd of shocked men and women. As she resolutely passed a sobbing Ruth Sokolsky, Baltz paused to say something, but the words didn't come. Looking toward the empty gate, the mother noted Milton La Riviere moving back to a seat beside the gate. As an attendant leaned over to say something to Mr. La Riviere, the father waved the woman away. "I am just going to sit here and wait on my daughter," he calmly explained. "I am just going to wait on Peggy. She will be here soon."

Not knowing what to do or where to go, a stunned Phyllis Baltz was on her way to her car when she was stopped by a United official.

"Mrs. Baltz," the man whispered, "they have found your son alive. He was taken to the Methodist Hospital in Brooklyn in a police car. We can take you to him."

The woman, who only minutes before had been forced to face having lost her son, didn't know how to respond. Could he really be alive?

Rushing by Baltz, a sobbing woman screamed out, "They found a boy alive. That's all! Everyone else is dead! Just one little boy!"

Stopping in her tracks, the shocked Baltz paused and said a short prayer of thanks, then grabbed her daughter and followed the man from United out into the dreary weather and a waiting cab.

Moments later, half a country away, William Baltz received an urgent call. Speechless he listened to the voice on the other end of the line, then drove straight to the airport and grabbed the next flight to New York. It was only after he had gotten airborne that the irony of having to fly to be with his son hit him.

In the days before live coverage on cable networks, most Americans kept up with the news of the midair collision through radio and newspaper updates. With each passing hour it seemed that another body was uncovered in the rubble in Brooklyn and Staten Island. Besides the passengers on the TWA and United planes, a dozen residents of Brooklyn were found dead under the twisted debris. Yet even as two schools and a church became temporary morgues and doctors were asking for dental records to help identify victims, the dead seemed quickly forgotten. From the White House to his hometown in Illinois, Stevie Baltz—the boy who had somehow lived through the nation's worst air disaster—was the center of each reporter's focus now. He was the miracle child.

With burns over sixty percent of his body, a broken leg and badly seared lungs, the eleven-year-old child was closer to death than life. Yet with eleven Brooklyn Methodist Hospital doctors assigned to his case, there was cause for hope. By evening he had even regained consciousness and was talking to his family.

"It looked like a picture out of a fairy book," Stevie said, remembering the scene out his window the second before the crash. "It was a beautiful sight. Then, all of the sudden, there was an explosion. The plane started to fall and people started to scream. That's all I remember until I woke up here."

As the night dragged on and his father and mother continued their bedside vigil, Stevie put the crash behind him and spoke of Christmas. He had decided that he wanted a portable television. After being assured that he would get one even before the twenty-fifth, the boy fell into a deep sleep.

Around the nation millions prayed for Stevie Baltz that night. The next morning, as he awoke seemingly stronger and more alert, there was reason for joy. Yet it was false hope. As time went on, young Baltz's lungs failed to respond to treatment and an infection began to quickly invade his body.

By his side, his parents and doctors urged the little boy to fight. They tried to make deals with God and took turns holding the child's small weak hands. Stevie responded to their words, he smiled and nodded his head, but his breathing grew more shallow with each passing moment. Finally he quit listening, turned his head to the side, closed his eyes and died. He was the final victim of two flights that spared no one.

In Chicago, at about the same time Stevie Baltz died, Sir Edmund Hillary, who had dodged death so many times as an adventurer, discovered he had escaped the grim reaper again. What he couldn't have known was that a lonely man was still waiting for the daughter who had gotten Hillary's ticket, and that a hundred other families were facing a mountain of holiday grief that seemed much higher and more impossible to climb than even Mt. Everest.

What happened in the air over New York City on December 16, 1960, was not only senseless and tragic, it was criminal. A crew flying without all of its navigation equipment in working order caused more than 150 innocent people to die. It cast a dark cloud over the holiday for New Yorkers, as well as scores of families whose lives were forever changed by the crash.

After the bodies were all found and trucks were brought in to haul away the tons of wreckage that littered New York, Stevie Baltz's body was placed on an airplane bound for Chicago. The plane's path would take Stevie over the exact spot where he had last seen the fairy-tale picture of New York from the air.

Meanwhile, the boy's family would not see the wreckage at Miller Field or the decimation in Brooklyn. Though United Airlines had offered them free first-class tickets, the Baltzes chose to ride a west-bound train. After they boarded, William Baltz pulled out a small sack that contained his son's personal belongings. There, in the middle of some loose change, a burned watch rested. Picking it up, Baltz held it in his hand remembering the way Stevie had constantly checked it as they had traveled to the airport. Turning it over, the father looked at the face. The watch's hands were forever frozen at 9:31, the moment the commercial jet crashed and the moment millions felt they were no longer always safe on a jet plane.

# 2

# The Galveston Hurricane 1900

According to weather experts there have been six more powerful U.S. hurricanes than the nameless storm that hit the southeast coast of Texas at the turn of the twentieth century, but for sheer death and mayhem, the Galveston Hurricane remains in a class by itself. In today's terminology it was *just* a 4 on a scale from 0 to 5, but even if it had topped the scale, somehow been measured at the impossible range of 5.5 or 6.0, it is doubtful that the Galveston Hurricane could have done any more damage than it did. It stacked bodies like wood and destroyed homes and businesses like a World War II bombing raid. It inflicted more pain in less time that any American storm of the modern age. And it was a storm that most said couldn't happen, at least not where it hit. Its killing spree was made possible by a misguided belief that if you lived in Galveston you were always safe from major storms. Based on Native American lore, oral history and blind optimism, this belief would set up an entire city for a death blow that still permeates every facet of Galveston's life today. The hurricane they called "the Great Storm" may have struck more than a century ago, but to listen to city officials talk, it could have been yesterday.

---

In the late summer of 1900, Galveston, Texas, was a community of thirty-eight thousand. This mecca for Gulf Coast trade had been built on sand, not rock, and the average elevation of the city was just four feet above sea level. The highest point in town was just over eight feet and that "hill" was man-made. Any engineer worth his salt would have warned those who continued to construct homes and businesses on the beachfront that disaster was just a crashing wave away, but with money generated by rapid growth, bank balances spoke much louder than common sense. No one warned anybody. The word on the street was that the city was immune to the forces that played havoc with the Mississippi and Florida coasts.

Before the Great Storm, Galveston was the crown jewel of the Gulf. For those barely above the grasp of the waves on a calm day, the city offered the finest of food, lodging, shopping and living, complete with an ocean view. Millions more people resided here than on any island in the country, outside of Manhattan. The mansions they built stood out as beacons of wealth to all the thousands of ships that sailed into the harbor. But just as the Texas hot spot was beginning to look like a rival for even New Orleans, a storm began to gather across the Atlantic off the African coast that would take the fine china, imported chandeliers, cases of expensive wines and everything else that gave Galveston the appearance of opulence and turn it all into rubbish. For the moment, though, there remained a lot to brag about and a lot of which to be proud.

With forty miles of streetcar lines, 2,028 telephones, and two automobiles, Galveston was a truly modern city during a period when most of Texas was still largely a wide-open frontier. The community had its own power plant, becoming the first Texas city to claim that every major building had electric lights. It had theaters and huge churches. It had an outstanding hospital and maybe the state's best school system. One obviously impressed eastern writer described Galveston as "a city of splendid homes and broad clean streets; a city of oleanders and roses and palms; a city of the finest churches, school buildings, and benevolent institutions in the South." In other words, to visitors and residents alike, Galveston seemed like a wonderful place—a heaven on earth. But with hurricanes in the Gulf of Mexico so common, and most people of the era deathly frightened of such storms, why was the city booming? Why was this seemingly tropical target growing at a faster rate than any city in the South? Ultimately it was all about rationalization rather than realization.

For over fifty years physical geographers had assured residents that the long, gentle slope of the adjacent sea bottom kept the city safe from the full fury of tropical storms. Storms would be broken up by the shallow shore long before they came on land. This meant that in terms of weather, Galveston was bulletproof. History seemed to have proven these experts right. While the city had been hit fairly often by supposedly major storms, they had become so weakened by the time they hit the island that little damage or few lost lives had resulted. So unlike residents in other port cities, the people of Galveston had grown very comfortable believing they were protected from the dangers of the sea. Most had even come to look forward to enjoying the sounds of a full-blown storm knocking at their door. The winds, rain and mild flooding had become a pleasant diversion. Folks from the Texas mainland even took the train to Galveston when word got out that storms were on the way. While a hurricane never brought destruction, it did bring a carnival atmosphere to the city.

It was on August 29 that a storm was born near Africa that would shake Galveston's security forever. By the thirty-first, that storm center had passed just south of Puerto Rico. On September 2, it crossed the Dominican Republic and Haiti. On the third, it moved mildly through Cuba. Finally, at 4 P.M. on September 4, the National Weather Bureau in Washington, D.C., issued its first telegraph advisory: "Tropical storm disturbance moving northward over the Gulf." That was all that the bureau said. Nothing more was needed. This storm had shown no real strength or passion.

Those at the local weather bureau, located on the fourth floor of the Levy Building in Galveston, read the report with a casual air. Cuban authorities had already informed them via a wire that this storm was not a danger. As a matter of fact, it seemed rather tame. Thus, no warnings were issued in Texas. Even if there had been, probably few in the city would have taken note.

On Wednesday, September 5, the storm raked the Florida Keys and transformed itself into a full-blown, if still relatively unaggressive, hurricane. In Galveston, with the temperature hovering in the nineties, few were worried. While the waves were swelling, residents had seen that signal before and nothing of note had happened on those past occasions. Ships were still docking, goods were still being transported and business was going on as usual for the men and women who lived in the third busiest port in the United States. Besides, they reasoned, perhaps a good shower would moderate the temperatures that had been scorching for most of the past week.

September 6 dawned like any other day for the people who lived in the city called the Queen of the Gulf. Children went to school, men went to work, and those in the industrial world began to tally the results of the first four days of the work week. A few wispy clouds passed overhead. Though the wind was howling from the north, bathers on the beach were enjoying the jumping surf. The warm waters hitting them as they waded added to their merriment and gave a bit of relief from the heat wave. Visitors to the shore, the thousands who regularly used the city's famed bathhouses, ate in the fine restaurants and checked into the luxury hotels, were hoping that the "good bathing conditions" would continue through the weekend, when the beach would be filled and a festival-like atmosphere would envelop the city. Yet as the evening activities took place as scheduled, by nightfall the folks at the weather bureau had finally started to take note of this latest Gulf storm.

The weather experts were concerned that the barometric pressure was falling rapidly, now down to 29.60. It had started out at a high of 30.29 (29.80 is considered normal) so as the pressure fell their concern increased. When it quit falling, they assumed the worst was over. Although they could not see anything on the horizon and the color of the skies didn't indicate that a major event was brewing, they decided to hoist a storm-warning flag above their office. The red-and-black banner went largely unnoticed by those who passed by. With the surf only one foot above normal and wind gusts topping out at just twenty miles per hour, most believed the warning was premature. Many saw the atmospheric conditions as typical of a warm summer evening, nothing more. Even those who worked for the Galveston Weather Bureau office had to admit that they were simply guessing that a hurricane might brush the city within the next few days. There was nothing concrete on which to base their initial alarm.

By Thursday evening at supper time, the temperature had dropped to eighty. It was still warm for late summer, but a good ten degrees cooler than the night before. With clear skies and a bright moon, the storm, if there was one out there, still appeared to be a long way off. After a week of record temperatures, a few were even wishing it would scrape the coast and cool off the oppressive heat. At the moment most were far more concerned about the prevalence of pesky mosquitoes than they were a hurricane. Besides, the barometer was now climbing, which meant to weather observers of the time that the storm was breaking up.

Friday seemed to prove that the flapping red-and-black flag was another forecasters' mistake. The surf was again pounding, but not threatening. Still it stood just one foot above normal. While the wind was blowing steadily at thirty miles per hour, making wading in the choppy waters a pretty miserable experience, a few swimmers braved an attempt. Still most bathers gave up before noon. Few bothered looking toward the horizon with trepidation. Instead, while waiting for the winds to calm, they took advantage of the shops and eateries.

The swell increased during the night and by morning several inches of seawater covered streets up to three blocks from the beach. Few cared and most greeted the expanded sea as an unexpected gift. Children played in the streams that ran along their streets. Bathers literally jumped out of the bathhouses into the surf, and streetcars ran double shifts trying to keep up with those who wanted to see the high tides. Few did much more than laugh at the inconvenience caused by the hurricane. Even those at the weather bureau agreed that there was really nothing to worry about. They predicted a few heavy rainstorms, but not enough to cause any real flooding problems.

By noon Saturday rain fell steadily. Even though the skies were gray and the water in the streets along the waterfront was rising, few worried. This type of minor flooding often occurred in cities by the shore. That is why almost all homes and businesses were built on stilts and why sidewalks were often four to six feet above street level. In spite of the inconvenience caused by the water, business went on as usual and tourists still made it to the beachfront businesses to shop. A few even swam.

While Galveston's citizens were innocently looking at this day as the start of the weekend, the National Weather Bureau in Washington was more concerned. Reports from ships indicated that this hurricane was a strong one and that it was bearing down on the east coast of Texas and west coast of Louisiana. About mid-morning, Washington telegraphed Galveston that while the storm's center was now expected to pass west of the city, there was still a chance that strong gail force winds and heavy rain might tear into the port as well. They thought the city should prepare for a major event and issue a major storm warning.

The locals who read the telegraphed warning smiled. With surf pounding and water rising, they had no doubt that a disturbance was just offshore, but they also knew that the long, shallow sloping bay would break up the hurricane's force long before it got to the city. Besides, even as the telegram had arrived, patches of blue were beginning to appear between the clouds. So rather than sound a major alarm, those at the weather bureau let things pass and the population continued to think of the intermittent rain that was falling in fits and starts as little more than a batch of summer showers.

By 12:30 P.M., the wind was blowing out of the northeast at more than thirty miles per hour. The barometric pressure was dropping and the rain was increasing in intensity. In the eastern and southern sections of the city, the water was now three feet deep. Some Galvestonians coming home for lunch had to wade through water up to their waists to reach their houses.

By 1 P.M., with the wind now at almost forty miles an hour, the streets and sidewalks within a few blocks of the coast had become full-blown water parks. Children were using washtubs as homemade rafts. The strong currents took them on bobbing rides along streets they had walked on the day before. Scores of others were jumping off the porches into muddy pools that used to be yards. It was a grand time! Even some adults joined in.

While the kids were laughing, horses pulling wagons and carriages strained as they made their way through water-swollen streets. With water up to their bellies, delivering passengers and goods was now a real chore. Streetcar operators were becoming frustrated as well. The water was washing the foundation away from the tracks, meaning that repair crews were going to have to come in the next week and fix the damage done by the rising water.

By mid-afternoon, the wind had reached more than seventy miles an hour, driving kids off the streets and bathers from the beach. Those walking on the city's high sidewalks were being blown to their destinations. On the beach, the surf was relentlessly pounding homes and businesses. And mounting piles of debris—baskets, driftwood and the like—were being driven through city blocks by the winds and waves. Even the streets in downtown Galveston were becoming streams. The water, racing rapidly from east to west, continued to rise slowly but steadily until about 6:30. Yet residents remained secure. They assured visitors that this sort of thing happened every few years and by tomorrow everything would be fine again. Even the experts agreed that the worst had come and would soon be gone. They joked that there would be no excuse to miss Sunday-morning church.

Those who had been to the beaches were not so sure. The huge bathhouses were taking a terrible beating. Porches were breaking off, as were slate shingles and shutters. Streetcar tracks were being washed away. Ships, though securely tied, were bobbing up and down like corks. The wind was now relentless, and gusts seemed to have the force of a locomotive at full throttle. Rain fell in fierce torrents and blew sideways. Some of the drops were so large and flying through the air at such great speed that they broke windows. Those on the beach looking at the rising water now doubted that this was just another storm.

On water-covered tracks, passengers on the Galveston, Houston and Henderson train were growing apprehensive as well. Even the crew was wondering if they would make it into the city. The train was inching along while men walked in front of the engine making sure that the track had not washed away and was safe for travel. Rain and wind pounded the cars, rocking them back and forth, and the water level, normally many feet below the tracks, was now above the wheels and threatening to come inside the passenger compartments. Worst of all, because of the pounding rain, the windows had to be closed, and with temperatures in the eighties, the sold-out train became a steam bath. Children were crying, claustrophobic women were fainting, and grown men were cursing. The minutes ticked by like hours, and all the while the water continued to rise and the skies grew darker. Finally, when the water rose and put out the engine's fire, all were forced to wade through waist-deep water for more than a mile to reach the station. Wet, tired and angry, these men, women and children were lucky. If the train had tried the crossing from the mainland during the evening, everybody aboard would have died.

A few hours later and with no relief in sight, the passengers sat shoulder to shoulder in the crowded depot. Many complained because no cabs would brave the weather to take them to their final destinations. Hungry and wet, no one seemed to guess that this was only the beginning. At this point most believed their discomfort was the fault of the railroad, not the weather.

To escape the humidity and heat that had invaded the old depot, a few men stood under the wooden porch and watched the swirling stream that only a few hours before had been a road. The water carried all kinds of things with it as it roared past them, and some began to catalog the parade of items. They laughed when they spotted an old hat, a tub, or a child's toy. The laughing stopped when one of them noted a dead child floating among the debris. It would not be the last victim that these men would encounter.

Another train was trapped on the other side of the bay. The ferry that was supposed to pick the cars up had been unable to dock. Most of the almost one hundred passengers stayed aboard the Southern Pacific passenger express, vowing to ride out the storm on the tracks. Nine frightened men and women got out and braved the elements to walk to shelter in a lighthouse. Those nine, jeered at when they left their cars, would be the only ones who lived to tell about what would soon be known as the Great Storm. The others would be drowned when the hurricane washed the tracks away.

At about sundown, some of Galveston's more religious citizens began to worry. It wasn't the weather that concerned them, it was the frogs. Suddenly, out of nowhere, tiny frogs not much bigger than a quarter or half dollar seemed to be sitting on everything that stuck out of the water. Twenty or more were fighting for space on a fence post while hundreds, sometimes thousands, were climbing onto high porches. This "plague" of frogs seemed to foreshadow something of biblical proportions. Mothers who had been laughing as their children played in the streets were now shooing them inside.

At the beach, the bathhouses were finished. The pounding waves had done in the structures. They had collapsed, as had the streetcar trestles and a number of wooden shops. As they fell, the waves had swept the debris into a damlike structure, keeping the surf largely confined in the beach area. Yet the wind and determined Gulf waters were now pushing this makeshift seawall a few feet farther inland each minute. With water levels more than ten feet higher on the bay side than the city side of that wall, if the mass of refuse broke up, Galveston would be hit by a wave that would cover almost everything in town. Yet no one knew it was there. No one guessed what was surely coming.

By now the phones at the weather bureau were ringing off the hook. Everyone wanted to know the latest word. None of the workers could tell the worried callers anything other than that the pressure was still falling and the winds had picked up to more than seventy miles an hour. Since the bureau's chief forecaster had gone home, seemingly unworried, this seemed to indicate that the worst must have already come and things would be calming in the near future.

It was true: Isaac Cline, the forecaster for Galveston, had gone home. It was also a fact that he didn't consider this storm a major threat. His story, brilliantly told in Erik Larson's 2000 Vintage book *Isaac's Storm,* played a major factor in the people of the city ignoring the hurricane until it was too late. For whatever reason, a lapse in judgment or ineptitude, Cline failed to identify the Great Storm's destructive power until it was too late. Those who trusted him would die along with many members of the forecaster's family.

While Cline was on his way home, most folks downtown had retreated indoors. At Ritter's Cafe a crowd had gathered for a supper. Though wind pounded the outside walls and rain peppered the roof, few noted either. Laughter rang out above the storm and drinks were hoisted all around the room. Then, without warning, the Ritter's roof was torn away, the foundation slipped to one side and the whole building collapsed. Three men died instantly, others were badly injured, and for the first time, the prominent citizens of Galveston began to realize what the frogs already had known—this was not just another summer storm. Finally those downtown, as well as those on the beachfront, knew that something awesome was out there just beyond their sight.

As darkness approached, the rain turned icy cold. With winds now above one hundred miles an hour, even those in the city's finest brick homes were growing concerned. In front of their houses, small buildings, work shacks, barrels, bottles, wagons and even animals were being dragged by the rising water that had now become a roaring river. The water was not fresh rainwater, either; it was seawater. The grave faces of those who tasted the salt on their lips signaled that the storm was now rattling Galveston's false sense of security to the core.

The water rose at a foot a minute until darkness, then the seawall made up of wreckage from the beach gave way. In an instant the water in the streets rose four feet. The Gulf was now ten feet above the street and even homes on stilts were taking in water on their first floors. Soon the power station lost its ability to produce electricity and the city was plunged into total darkness. Wading through water that in some cases reached their heads, families rushed out into the streets to find shelter in structures they thought stable enough to withstand the storm's power. Some drowned in the panic-induced dash for safety, others would live long enough to see their havens of security smashed by the winds and waves.

During the next hour, the water would rise another five feet. Bodies were everywhere. Children, some crying for help, were caught up in the streams that had been their arena of adventure a few hours before. Mothers jumping into the rapids drowned trying to save them. Fathers, many caught downtown and unable to get home, cried at the thought of their families facing this monster by themselves.

The homes that withstood the rain and wind were now being deluged with seawater. First floors filled with water, then second floors became pools of death. Soon the only thing showing were roofs. Those who stuck with their homes died. Those who jumped had to face hours in the raging surf, riding around the city on planks, pieces of broken-up homes and barrels being pelted by the raging wind and water and by thousands of other airborne missiles.

Three miles west of the city, the rambling wooden buildings of St. Mary's Orphanage stood on the beach. The lonely structures housed ten Roman Catholic Sisters of the Incarnate Word and ninety-three orphans. The sisters, convinced this storm was a killer, took all the children to the chapel on the first floor of the girls' building. They stayed there praying until rapidly rising water drove them to the second floor. From there they watched the boys' building break up in the storm. About an hour later the roof of their own building collapsed. With death threatening from every side, their schoolyard now covered by almost twenty feet of angry water, the sisters tied the children together in groups, attaching ropes to their own waists and then to the walls. Three older boys remained free of the ropes in order to serve as lookouts. With winds now raging at more than 150 miles an hour, the dormitory could no longer take the pounding. As if hit by a bomb, it exploded. Torrents of water rushed in. With wood splintering all around them, the children and nuns were caught up in the current. They had no chance. The rising surf quickly covered them with seawater. Tied to the stronger parts of the building, most drowned or were beaten to death by debris before the walls gave way. The few who did float in groups to the surface had no better chance. Death simply took a little longer. The ropes that were meant to save them got tangled in wreckage and pulled them back under the water to die together. Those ropes would hold on to their bodies until they were found days later. The only three to survive would be the boys who had served as lookouts.

Hundreds who had been driven from their homes were now being killed by the most innocuous things. Toys, fountain pens and small garden tools were being picked up by the wind and were finding human targets. Wood, bricks and pieces of glass had also become bullets. But the most efficient man-made killer turned out to be a part of the city building code.

All homes in Galveston were required to have slate shingles. This was because a fire that had raced through the city two decades before had been spread by wooden roofs. With the winds now ripping the shingles from homes, the very thing meant to save lives was taking them. Pieces of slate were flying through the air as though fired from a cannon. People were being beheaded by them as they tried to flee the surging water. As if the storm, with its wind and rain, was not enough to contend with, the hurricane now had an arsenal of lethal weapons at its disposal as it raced from block to block.

Though there was nothing left to measure their intensity, researchers estimate that by midnight the winds were in excess of two hundred miles an hour. Rain, blowing sideways, was falling at the rate of several inches an hour and the sea was now covering the whole island. Only multistory buildings were sticking out above the waves, and they were being rocked by the rivers of debris and water that raced by them on every side.

Thousands of Galveston's finest were in those streams. Some were already dead, a few were crying for help, many were desperately clinging to pieces of roofs or porches. Fathers and mothers were frantically searching through the night for children who had been beside them a few moments before and then had silently slipped off into the water. Flying slate tiles were still killing scores, and walls of debris, some more than fifty feet high, were pushing through town like bulldozers, destroying everything they encountered. The storm was so loud that even a ship's horn could not have been heard a few feet away.

As Galveston was literally washed away, the barometric pressure had dropped to 27.49. It was the lowest reading ever recorded in the United States and it was destined to fall even more. Today it is believed to have touched 26.7. Being hit from both sides of the island at once, Galveston had no chance. Ships in the harbor were being dragged back and forth by the tides, their moorings gone. Like the city they were also at the mercy of the storm. Everyone was a victim and no one would survive the long night without suffering a loss.

In the midst of the pelting rain and the raging tides, something surreal lit up the sky. A full moon shown through the clouds. With terror reigning and thousands riding out the destruction in the water, the moon bathed the scene in an eerie light. It was then that those who had once thought this nothing more than a summer storm were hit by a numbing fact. There had been no lightning or thunder, no real signs that the greatest storm to ever strike the United States was anything more than just another gale. With the moon shining, the Great Storm was still disguising itself in sheep's clothing.

If the night had been terrifying, the dawn brought horror. All that remained of thousands of homes and of countless people was a three-mile-long mound of wreckage jammed with bodies forming a semicircle around the business district. Outside this arc not a building was standing, not a street was outlined. Wind and water had restored the land to primal beach.

Those who boated over from the mainland were horrified. The water was filled with corpses. They knocked against the sides of the boats, bobbing like driftwood for as far as the eye could see. When they arrived at Galveston, few believed that anyone could have survived. The fact that one in five Galvestonians somehow made it through the terror-filled night is a miracle.

On the ground, men, women and children cried out for their loved ones. They searched the faces of the dead and pulled away rubbish looking for someone that they had once known. They prayed, they cried, and they went back to the places where they had once lived. With nothing left to call home, they wandered, looking for anything that provided a glimpse of the life they had lived until a few hours before. Some found a picture, a book, a toy or piece of broken china. Most found nothing that they recognized. Thousands were in shock, no one could fathom what had transpired just hours before.

First reports, estimated by a local priest, guessed the death total at a staggering five hundred. Within days that number had grown to six thousand men, women, and children. The final number was closer to eight to ten thousand. It would take more than thirty days to search through the wreckage for the bodies that were not carried out to sea. Mass graves were dug, funeral pyres were lit and the act of getting rid of the dead became an almost methodical routine, devoid of feeling. After a while, the bodies were treated in almost the same fashion as the debris. Stack and burn, stack and burn, stack and burn. Mourning gave way to silent submission.

Besides those it killed, in the end the Great Storm left a thousand survivors completely naked and five thousand more bruised and battered. In a fifteen-hundred-acre area of total destruction, 2,636 houses, nearly half the homes in the city, were swept out of existence. Elsewhere, at least a thousand more structures were reduced to wreckage. Not a single building escaped damage. An estimated eight to ten thousand survivors, almost a third of the population, were homeless. For several weeks the bay was littered with hundreds of human bodies, the corpses of cows, horses, chickens, and dogs.

Galveston recovered and rebuilt, but the city would never again rival New Orleans or Baltimore in size or stature. Houston would even take its place as the port city of Texas. Today the island is as much about what happened in 1900 as it is what is going on today. Never again will the people of the city take a hurricane for granted. Now warnings go up early and most heed them. When the storms approach, citizens head to the mainland. While the likes of the Great Storm has not been seen for a century, those in Galveston will never feel completely safe. They know that one like it could and will return, they just don't know when.

# 3

## The New London School Explosion 1937

Sometimes explosions are caused by chain reactions. Often a series of different events occur, resulting in the final blast. Usually these events play out in minutes or seconds, but in the case of the New London school, it took years for a series of seemingly unrelated events to set off an explosion that would rock the whole world and kill almost three hundred unsuspecting people.

New London was a sleepy East Texas hamlet when the stock market crashed. Most of the folks who lived in the small, out-of-the-way Rusk County community were hardworking people who made what little they could from the ground. Farming hadn't offered much of a living before the bottom had dropped out of the market. Now, in the midst of the greatest depression to ever grip the nation, those who lived around New London had even less hope. Yet one event would soon change that and make this area one of the richest in the world.

C. J. Joiner, a nearly seventy-year-old Oklahoman, delivered to Rusk County the key for escaping the Depression. A lifetime oil man, he sensed that under the sandy Texas soil were huge pools of black gold. He drilled two wells in the late twenties. When neither of them produced, people shook their heads and urged him to move on. Undaunted, he tried again. On May 8, 1929, Joiner hit a gusher. Within two years, the boom was on.

Seemingly overnight, oil derricks were built almost on top of each other. You could find them in the middle of rural roads, on city streets, in front of businesses and in schoolyards and churchyards. And New London, a town the Depression should have sucked the life from, was experiencing sudden growth and prosperity like it had never known.

With the influx of new people and business, New London hardly noticed that most of the nation was living in the poorhouse. Folks here didn't need handouts, the WPA, CCC or any of the other FDR-sponsored government programs. Even as the rest of America was starving, life in Rusk County was good and opportunities seemed endless. By 1936, Joiner and others had hit so much oil that thousands of people from all over the country had been brought in to man the jobs created by the massive discovery. The oil boom had created a population boom!

Hundreds of men flooding into the area to work in the oil fields meant thousands of new children too. The modest buildings that had once housed schools simply were not big enough anymore. With tax coffers bursting at the seams, many of Rusk County's towns began to construct new schools and New London followed suit.

The New London school board reviewed plans from all over the nation before choosing one to house their junior high and high school students. The facility that oil money built was quickly labeled the finest in the state of Texas. Newspapers and magazines featured the beautiful two-story structure, hailing it as a model for all American education. Almost every one of these stories also labeled the New London, Texas, Independent School District as the richest rural school district in the United States.

The high school was a beautiful, modern, steel-framed, E-shaped building on the outskirts of own. It featured an incredible auditorium, 53 feet wide and 107 feet deep, that seated seven hundred. It housed seven main classrooms, a manual training room, science laboratories, executive offices, a drawing room and rest rooms. It was fireproof, with steel bracings instead of wood studs used in the walls and the roof. Every facet was state of the art. Teachers, students and townsfolk considered it the pride of New London. They bragged that oil might have put the town on the map, but it was education that would make a more lasting impression on the world.

On March 18, 1937, the New London High School was preparing for a regional academics tournament in nearby Henderson. Naturally the school was caught up in great excitement as teachers and students readied themselves for the competition scheduled for the following day. The hallways and classrooms were alive with the buzz of motivation and hopefulness. After weeks of work the students thought they were prepared for anything. They believed that they were going to bring the top prize back to New London. As it turned out, they would not get the chance to answer a single question.

Early in the week there had been some talk of dismissing the students at 2:00 P.M. rather than the usual 3:15 P.M. on that Thursday. Yet the need for last-minute drilling to ready students for the contests, as well as accommodating a previously scheduled PTA meeting, nixed those plans. Needless to say, the students were disappointed. They had hoped to catch an early movie, listen to some music or do some shopping. Now it was simply school as usual.

The eighteenth of March was a fairly typical spring day, with light breezes and the sky a mix of clouds and sunshine. The school's heater was in operation, but some students wore coats and sweaters as they walked down the halls—many of the classrooms had their windows wide open. It seemed that while much of the new school was state of the art, the steam heat was not. Powered by natural gas, it was often cranky, unpredictable and unstable. On warm days it worked too well, on cold days it often didn't seem to work at all. Some teachers, whose students complained, had suggested that some gas was escaping into their rooms and causing burning eyes and headaches. It was these teachers who usually threw their windows open even on the coldest days. Yet no one sounded an alarm that the situation had to be fixed. Warm days were just ahead so the heat would soon be turned off anyway. If there was a problem, then the maintenance crew could fix it during the summer break.

Besides the staff and students, most of the parents realized that gas was leaking in some areas of the new building. Yet in the middle of oil country, this seemed to cause little concern. Surely, most reasoned, if anything was really amiss, it would have already been caught.

Earlier in the year, "the richest rural district in the country" had even decided to rid itself of its three-hundred-dollars-a-month gas bill and switch to a "free service." "Save anywhere you can" seemed to be the school board's motto as they told the community about their prudent move to cut utility costs.

Parade Oil, located just across the street from the campus, had offered to let New London use its refuse natural gas to heat their buildings. This raw or "green" gas, pulled from the oil wells, was usually just burned at the pump. It was not worth enough to justify refining, when oil, the real moneymaker, was available in such huge quantities. So the common thinking at the time was to give it to anyone who could use it. Certainly the school could benefit from this service.

The school board was not going to look a gift horse in the mouth. The monthly savings by switching to green gas could buy a lot of books, desks and sports uniforms. Yet there was a problem with accepting this gas. Unlike commercially purchased fuel, what the New London school was using had no smell. Hence, there was no real way of knowing when there was a major gas leak. Also, the gas pressure from this system varied widely from day to day. On some days there would be barely enough pressure to fuel the boiler. On others, the pipes were so filled and the pressure so great that the gas sought out any leak as a release. These were the days that the furnace worked great, but the headaches and eye problems were most severe.

Today, few parents would allow their children to go into any building fueled by raw gas. In the 1930s, a host of homes in Rusk

County tapped into this source for heating, hot water tanks, stoves and clothes dryers. Few feared it, and using raw gas was seen as a way to fully take advantage of the gift of living in the midst of an oil boom.

About an hour before school was to be dismissed, scores of parents, mainly mothers, began arriving in front of the new building. Several of the PTA members walked into the high school thinking that their scheduled meeting would be held in the auditorium. It usually was, but the March assembly had been moved to the gymnasium on a different part of the campus because an elementary group was going to perform a Mexican hat dance. Carefully walking across the soft wet lawn—the area in front of the high school had just been sodded—the parents assembled on the bleachers, listened to a prayer and watched the enthusiastic performance. Except for the noise of tiny feet on the hardwood floor, the melodious strains of musical instruments coming from the band hall and the sound of some boys practicing on the baseball field, nothing seemed unusual. To most it was just another day.

At five minutes after three, teachers and students grew restless. It was time to wrap up the day. Superintendent Shaw left the main building to walk across campus for an appointment. He was smiling and waving at the parents who were beginning to arrive to pick up their children. In Shaw's mind he had a great job and the people of New London were some of the best he had ever known. He couldn't imagine ever wanting to be anywhere else or do anything else.

As Shaw casually strolled down a sidewalk, Della Westbrook, a teacher, decided that she was hungry. She didn't have a class last period and opted to go across the street from the high school to a small food stand. At about the same time that Superintendent Shaw had walked some fifty yards from the high school's front door, Westbrook was staring at a menu trying to decide what would satisfy her cravings.

With just five minutes to go before the bell, Lemmie R. Butler, the shop teacher, was working with his class on a project. He picked up a piece of wood and explained why and how it needed to be sanded. As he spoke he had no idea that his eyes were watering and his head pounding because his room was filled with natural gas. He also couldn't have guessed that the boiler room and all the school's walls were filled with the silent lurking monster as well. Neither Lemmie nor anyone else had any reason to believe that the pressure coming through the ill-fitted gas lines was so high that more than sixty-four thousand cubic feet of gas had escaped into the building. If it had been commercial gas, the smell would have been overpowering. Yet with green gas there was no smell and no signs other than the dripping eyes and pounding heads.

After explaining the next step of his lesson, Butler reached over and flipped the switch on a power sander. It would be the last lesson he ever taught and the last action he ever took.

For some reason, Superintendent Shaw turned back toward the high school at the same moment Butler flipped the switch. Just as he did, a wave of turbulence knocked him off his feet. Across the street at the food stand, Della Westbrook had also turned to admire the beautiful building. Just as she did, she felt the ground shake. Managing to stay on her feet, she watched in shock as the building's roof lifted from its walls and hovered in the air just long enough for the walls to blow outward. A split second later, the tiled roof was crashing downward, not stopping until it met the ground. In the blink of an eye, the carefully constructed 253-foot-long building was gone.

Mrs. Homer Gray, the study hall teacher, heard the blast and felt the building shaking. Not knowing what was happening, she ordered her students to get under their desks. Plaster began to fall and pieces of the ceiling started raining down on students' heads. Some of Gray's charges panicked. Screaming, they pushed windows open and jumped to the ground some fifteen feet below. As they did, the high school crumbled around them. The rest were buried in tons of rubble.

One shocked witness, almost a football field away from the school, looked on in unbelieving horror as the school collapsed before his eyes. His children were in that building. What had happened to them? He didn't notice a two-ton concrete slab hurtling through the air toward his car. He didn't have time to even turn in his seat before the block crushed the Ford with him inside.

Superintendent Shaw pulled himself to his feet even as the tiles and chunks of stone were raining down around the building. As he peered through the dust, he saw the bodies of scores of children. They had literally been blown through walls and windows. Some were lying dead only a few feet from him, while others were crying out in pain.

Della Westbrook forgot about her hunger and began to race back across the street, dodging debris that was flying at her from every direction in order to get back to what was left of the school. As a fine mist of dust began to settle over the ground, bricks flew past her head, as did body parts of some of the students she had taught only an hour before.

Those closest to the center of the blast, such as Shaw and Westbrook, didn't actually hear the explosion. They just felt it. But in the gymnasium, the PTA was rocked by a deafening roar that echoed off the walls, ceilings and floors. The bleachers rattled, lights swayed and plaster fell from the walls. The explosion hit many so hard that they were knocked off their seats. Some blacked out. In the seconds that followed, as they pulled themselves up or came to, the parents realized that something terrible had just happened to their children. A white dust drifted everywhere and an awful silence had replaced the sounds of a normal school day.

Those who witnessed the explosion said that there was a quick sheet of flame that raced through the ruins right as the dust began to settle. If that was true, then the school was genuinely fireproof, because there was no evidence of fire among the twisted girders, broken stone, and shattered tile. That there was nothing left to greet those who struggled out of the gym and raced over to the high school was just as disturbing. With the exception of two sections of a back wall, the rest of the structure was gone.

Two miles away, employees working in the Humble Oil Company offices were checking on reports and getting ready for a coffee break when they felt their building rock and heard a huge blast. "Oil well!" most assumed as they hurried outside. Yet the cloud of smoke and dust was not hanging over a derrick; it was drifting away from the school grounds. Most immediately jumped into their cars and headed for the scene as offices were left empty and work forgotten. The same scenario was repeated all over New London.

The debris field stretched for almost a half mile. It contained not only building materials, but books, lockers, coats, shoes and even pieces of bodies. Many citizens, driven out of their homes by the blast, found everything from lesson plans to sports equipment littering their yards and the streets. For several minutes paper floated down like snow and white dust hovered in the air.

Back at the school, children and adults were beginning to take stock of what had happened. As soon as they felt the earthquake-like rumblings, Milton Harris and Geraldine Ruffin had abandoned their second-floor classroom and jumped to the fresh, soft sod. Neither was hurt, but both were in shock as they observed classmates and friends thrown through the air like rag dolls.

Evelyn Peters didn't have time to consider jumping; she was blown through a window. Though bruised and disoriented, she was unhurt. Not far from Peters lay Bernice Morris. Morris had been on the ground floor in a drawing class; the blast tossed her out through a window. Like Peters, she was relatively unhurt. A few feet away, J. B. Nelson also found himself on the ground. He too had been lucky enough to be sitting next to a window when the blast hit. Most were not as fortunate.

Scores of students were pushed by the force of the explosion into walls. Many others were decapitated by flying sheets of glass. Countless young people were smashed by stones and tiles. Almost all were buried under rubble.

Several parents, waiting to pick up children, didn't have time to react to the blast. Huge stones sailed through the air like missiles, some tearing through cars, others flattening unprepared people. The lucky ones had been knocked to the ground. Those who managed to maintain their balance often found themselves directly in the path of huge pieces of the school flying through the air at hundreds of miles an hour.

The school's plumber was one of those dodging debris. He had been working on a bathroom fixture just moments before. His life was saved by the need for a tool. He had gone outside to his truck at just the right moment. Now, like scores of others, he knew he had to get back into the rubble to try to save lives.

The students who were alive began to fight to free themselves from the wreckage. Evelyn Hudkins, just sixteen, was buried under piles of bricks and tiles. Pushing herself free, she felt for her boyfriend, Johnny Roberts. He had been sitting beside her. Grabbing his arm, she crawled through twisted wreckage, all the while pleading with him to help her find a path to safety. He didn't respond. Hudkins continued to pull away stones, broken desks and tiles until she fell out onto the yard. The girl then reached back and dragged Roberts to the schoolyard too. When she got him outside, she realized that Johnny was dead.

One of the few witnesses who saw the actual blast was driving a car several miles away. She told the *Dallas Morning News* that she felt a tremor shake her car, then saw the smoke rise in the air. She heard the rumble of what sounded like thunder. The smoke, which was actually dust, covered the derricks and reached for the sky like a giant mushroom. It was only then that she saw objects flying out of the dust.

As men, women and children poured out of the elementary school—which was undamaged though only a hundred feet from the high school—band hall, gymnasium, and nearby homes and businesses, they were greeted by a bizarre scene. A layer of silt already covered everything and everyone who had been close to the high school. Through the foggy, powdery air, witnesses could barely make out children who had been blown out onto the lawn.

For a few seconds, everything stood still. Most of those who had not been in the school were too stunned to move. While many immediately realized what must have happened—that there had been a mighty explosion—and while some even guessed that it had been triggered by the natural gas used in heating the high school, few fully realized just how powerful that blast had been. The evidence was visible to them, but it was incomprehensible as well. Their children had been inside those walls. They had to be in there still. But if this was the case—most refused to accept the reality of what any logical soul would have guessed—how could any of them have survived?

As more people raced up to the school, hysteria set in. There was a feeling of helplessness that hovered over the area as thick as the debris mist and just as confusing. As parents, teachers, the superintendent, other students and neighbors pushed their way closer to the building, they began to hear prayers, moans and cries for help. They also began to make out students who had either been blown out of the building or had somehow crawled from under the fallen walls and ceiling in the moments after the school had been leveled.

Within minutes those who lived or worked nearby began to arrive. The dust was still settling when R. K. Carr arrived at the building. He ran across the yard and immediately began searching through the rubble. He was pulling away stones, books, lockers and tiles, frantically looking for any signs of life. Hearing a cry, he pushed away a mound of debris, reached down and grabbed a dust-covered small girl. Lifting her into his arms, he discovered that it was his own daughter. Crying tears of joy, he carried the child out to the safety of his car.

George Hardy also raced to the ruins. Yet George didn't see any signs of life; he saw only the dead bodies of children. In his sixty years he had never even imagined such horror. Tears clouded his eyes as he desperately searched for someone to save, only to be greeted by the cold stares of the innocent dead time and time again. He fell to his knees, grabbed his head with his hands, and died of a heart attack.

A mother, dressed in her finest outfit, ran from the PTA meeting across the lawn to the school and yelled out for someone to call for help. Pushing her way past hysterical parents, she worked her way to the school. All around her children were reaching out and begging for help. It was almost as if she didn't see them. She was lost in a fog, urgently looking for a dress or coat that would signal that her daughter was not buried under the building. As a stunned girl with a deep gash in her head came stumbling out of the mist, the mother grabbed her. Not noticing the child's injury, the woman cried out, "Where is Betty, where is my daughter?" That mother's pleas were quickly echoed by a dozen others. In their panic, parents were trying to get to the rubble to uncover their own, often ignoring other wounded or dead children whom they had known for years.

A salesman who had arrived at the school a half hour late for a meeting parked his sedan amid the rubble and slowly walked toward the scene. He counted a dozen or more dead children lying in the yard. Although many of them appeared unhurt, they weren't moving. They had probably been killed by the force of the blast. Others were missing limbs, but had died instantly, so their terrible injuries hadn't produced much blood with all the gore. As the man surveyed the human toll, he felt as if he were watching a movie or walking through a nightmare, but this was real, so real that he fell to the ground and wretched.

For almost fifteen minutes there was no real organized rescue effort. In many cases, mothers and fathers dug through twisted metal and bits of block with bloody bare hands looking for loved ones and ignoring others. Teachers who had not been killed or injured were digging too, but they were just looking for anyone who was breathing.

As time dragged on and more and more dead bodies were found, parents went into shock. Disheveled, crying mothers often fought over injured or dead children. Arguments broke out over whose child had just been found. All reason was lost as men and women would claim a red-headed girl as their own when their child had blond hair. In those first moments there was little logic, only grief and anguish as the body count began to grow.

As delirious mothers and fathers cried out their loved ones' names, a Humble Oil representative drove up. Seeing the massive damage, he sent immediate word to round up every roughneck that was in the field. He also asked them to bring the oil company trucks, bulldozers and all types of heavy equipment. This call was the beginning of what would become a massive rescue effort.

By 3:35, Humble's first trucks had appeared along with more than a hundred men from the company's fields. They would soon be joined by nine hundred others from a host of other oil companies. Using every piece of equipment in their arsenal, these men put together an organized effort to move debris and get to injured students and teachers.

Within an hour of the blast, Governor James Allred was notified. When he was told of the chaos that was enveloping New London, he ordered the Texas Rangers and highway patrol to the scene. Allred also put his staff to work assembling a team of doctors and nurses to aid victims. Hospitals in Dallas and Tyler were called and medical supplies and personnel were rushed to New London. Within two hours of the blast, aid began arriving from Baylor Hospital and Scottish Rite Hospital for Crippled Children in Dallas, as well as from Nacogdoches, Wichita Falls, and the United States Army Air Corps at Barksdale Field in Shreveport, Louisiana. These professionals were assisted by deputy sheriffs from Overton, Henderson, and Kilgore, by the Boy Scouts, the American Legion, the American Red Cross, the Salvation Army, and volunteers from the Humble Oil Company, Gulf Pipe Line, Sinclair, and the International–Great Northern Railroad. Even the Civilian Conservation Corps and their trucks were moved to the school. Yet there was little that this massive army of trained professionals and volunteers could do. There were few left to save. Worse yet, by nightfall it was raining. Still, after finding more than a hundred dead bodies, no one gave up.

Working in a steady rain, frantic men and women dug through tons of rubble. When darkness came, Humble Oil brought in a generator and a bank of powerful lights. Rescue operations continued through the night. Lit by the blue-green light of acetylene torches, the workers, sweating and shirtless, labored past collapse. Many were driven beyond reason because they were looking for their own children. And as fewer and fewer bodies uncovered showed any signs of life, the hopelessness of the situation could be read in their eyes.

By midnight, those found in the ruins looked more like battered dolls than humans. They had been buried under tons of rubble and had probably died instantly. Yet the men were pushed on by the hope that maybe one more child could be found alive. As body after body was dug out and carried off to the school gym, the work continued.

A young Houston news reporter on his first major assignment, Walter Cronkite scribbled down bits and pieces of information as he watched the rescue effort. Cronkite was amazed by not only the destruction at the site but by the way volunteers worked in an effort to find just one more miracle. The reporter, whose job called for him to remain objective and distant, couldn't help but feel the pain of men like Joe Davidson. Davidson, a World War I hero, had been digging along with scores of others for hours only to uncover the sad truth that his two daughters and sons had died. Like so many others, Davidson was too shocked to even grieve. He simply told those who would listen that his son had been planning on going to the Naval Academy and then went back to work looking for others who might somehow still be alive.

As the now almost hopeless rescue work continued, many parents talked about their children's dreams. "You know she wanted to be a teacher," a mother sighed. Another added, "He was going to go to the University of Texas in the fall." Holding on to what they could of a life that had been so whole only a few hours before bought time and solace, things that were in short supply now in New London.

At 8:00 P.M. hope was rekindled as the workers announced they had found someone alive. Digging through the rubble, huge arms pulled Naomi Bunting out of the debris. Naomi grabbed a hand that reached for hers as she breathed clean air for the first time in hours. Rescuers paused and smiled. Mothers clung to reborn faith. Yet as the eighteen-year-old was placed in the ambulance, she died. They had found her too late.

Mrs. Gray, the teacher who had saved her students in study hall, was found at three in the morning under her desk. She was alive, uninjured but stiff. Her students were dug out too; all of those in Gray's charge lived.

Just before dawn, another teacher, Tracy Tate, was pulled out alive. But as she was placed on a litter, she sighed and the life left her body. A thousand men working as fast as they could were not fast enough for Tate or scores of others.

A few minutes later a worker found another miracle. This one was not human, and could have snuffed out even more lives. A box of dynamite, left over from explosives used in blasting away the rock around the new football field, had been stored in the school. Somehow it had not exploded with the building or when the cutting torches slashed through material all around it.

Dawn's light brought the reality of the situation into sharp focus. Most of the building had now been excavated. There were few places left that could hold intact bodies. It was time to identify those who had been found. As lawmen kept parents away, teachers were asked to look at each body or body part to try to make identification. Some students, those who had not been disfigured in the explosion, were easy to spot, but others looked like total strangers. In many cases there simply was not enough left of a student or teacher to identify. At that time, numb voices would call out, "A small boy, red shirt and blond hair," then wait for someone to come forward and claim the body.

Churches became morgues, children who had miraculously lived had been taken to hospitals in every direction. Cars, mostly driven by the curious who had traveled to see the ruins, choked highways. Tempers flared as parents looking for missing children tried to get to temporary morgues or to hospitals in Henderson, Overton and Tyler, and were caught in the traffic jams. When they finally arrived at the medical centers, they were rarely met with good news.

Newspapers across the country reported that workers had uncovered more than 650 dead children. The stories spoke of the need for medicines, doctors, nurses, embalmers and caskets. On Friday morning cameramen shot pictures of rescue workers using peach baskets to pick up body parts. And hundreds of reporters watched as mothers and fathers tried to piece together their shattered families.

Lama Stround was one of those the press followed. Reporters had listened to her pray as she looked at body after body for her own child. They were there on Friday, staring into her stricken eyes as she viewed the lifeless form of her sixteen-year-old daughter, Helen. And they watched as the woman fell to her knees in anguish, tears streaming down her face, moaning in pain, before collapsing to the ground, dead.

For Helen and three hundred others, Dallas coffin makers worked overtime. Those who were going to be conducting funerals began to round up hearses and grave diggers. For those selling caskets and flowers, business was never better.

As the story hit the world's newspapers, people responded. Ironically, the first telegram of condolences from a foreign leader was sent by Adolf Hitler. The man who would use gas to kill millions was somehow moved by the disaster.

In the midst of the incredible sadness, a few fortunate families celebrated. G. O. Sanders could have lost three children. Yet his oldest daughter, Vitra, had left school earlier in the day to take a shorthand exam in Henderson. His middle child, Glen, was playing baseball behind the gym and his youngest, Robert, was at home ill. The Sanders family miraculously lost no one.

Robert Norris had been at school that morning, but had decided to play hooky after lunch. It was his birthday and he didn't want to get spanked by his friends. When he left home, he went to a movie. He was shocked to come out of the show and find that hundreds of his friends had died.

Robert Williams had snuck out of class during the last period and escaped the cataclysm. And Lee Marsden, Luther McClure and Truman Honeycutt had cut school to go to the cattle show in Fort Worth. These decisions, which would have been viewed as immature and rebellious, had in this case saved each of their lives.

By Sunday the schoolyard had been cleared and the real story of grief and suffering was being lived by thousands. Though the newspapers had gotten it wrong and only around three hundred had died (the school had only just over five hundred students enrolled in the spring of 1937), funerals still had to be held like relay races. One followed another, and as many as five or six caskets were in the church for each service. Graves were dug twenty-four hours a day and burials often had to wait for diggers to finish. Seventy-five children were buried at Pleasant Hill Cemetery alone. Because there were not enough hearses, trucks and wagons were used to transport the bodies up and down the rain-soaked muddy roads. The press and the curious dogged the mourners every step of the way. To many this was the ultimate sideshow.

Of the 500 students and 40 teachers in the building, 298 died. Only 130 students escaped serious injury. The dead included 16 teachers, 4 visitors, 63 fifth-graders, 79 sixth-graders, 33 seventh-graders, 29 eighth-graders, 9 ninth-graders, 14 tenth-graders, 19 eleventh-graders, and 4 more who were taking additional graduate-level courses (grade eleven was as high as most Texas high schools went at the time). There were more who would have been included in the final breakdown, but their families chose not to have their names listed.

By March 21, 1937, as the debris was picked up and hauled away, a chalkboard was found in the rubble. The writing on that board was as clear as it had been the day of the explosion. It read, "Oil and natural gas are East Texas' greatest mineral blessings. Without them this school would not be here and none of us would be here learning our lessons." One incredible lesson was learned just moments after those words were written. Unfortunately, it was too late for three-fifths of those who studied inside the building built by the richest rural school district in the world.

# 4

# The Great Nashville Train Wreck 1918

In 1918, for the first time in history, the United States sent young men in large numbers to another continent to fight. With a world war raging in Europe, the doughboys were answering the call to go "over there" to win the battle for free men everywhere. As they geared up to leave, many of the inexperienced soldiers, as well as the families they were leaving behind, were very nervous. This was a new and frightening experience for the nation's people, and the security of being an ocean away from the warring European nations had now been shattered. No longer did being on the other side of the world mean that Americans were safe. By this time, German U-boats had successfully debunked the concept that the U.S. was an entity by itself and not connected to the rest of world.

It was hot and clear on July 9 of that year, and Nashville's Union Station was filled with young men who were anxiously looking to a day in the not-too-distant future when they would lay their lives on the line in faraway France. Many, away from home for the very first time, were facing an adventure that scared them to death. Some, who had come from farms in the deep south, had never even ridden a train, much less seen the ocean or been on a ship. This was not a day they relished, it was a day they dreaded. The train that waited to take them away from the secure life of their youth was one they didn't want to board. As they saw it sitting beside the station, many prayed that train Number 4 would somehow simply disappear, letting them resume their normal lives.

Among the hundreds of soldiers that early morning were many young African-American men who had been recruited from the farms around Nashville. The promise of money had lured them from a fairly sheltered environment into a world that was new and often hostile to them. They had signed up to work in munitions plants and they knew that their jobs were probably as dangerous as fighting on the front lines in Europe. In many ways, the train that was waiting for them had death written on it as well.

As they milled through the station, both the workers and the soldiers, though separated by a river of racial prejudice, had a lot in common. They all wondered if they would be lucky enough to ever ride a train back to Nashville, or if the ticket they held was a one-way passage to glory. Little wonder that many of them stalled as long as they could before finally boarding the cars and finding a seat.

In front of the anxious men who were slowly filing onto the train stood David Kennedy, "Uncle Dave" to those who knew him well. For several minutes before finally jumping into the cab of engine 282, he watched soldiers and civilians climb onto his train. After checking his watch, the old engineer, who had more than thirty years' experience with the Chattanooga and St. Louis Railway, glanced over at his fireman, thirty-four-year-old John Kelley. Kelley could see in Kennedy's eyes that the engineer was deeply frustrated. It bothered both of them that the train was already more than five minutes late leaving the station. As the fireman got the boiler ready, Kennedy kept checking his watch and impatiently shaking his head. "Get them loaded up," he said to no one in particular, then glanced back as a few more stragglers ran out of Union Station toward the train.

Kennedy was the NC& St.L's senior engineer. Though today it is difficult to imagine, he was also a local legend, the top player in a world where men like him were idolized more than baseball players or politicians. When he walked through the station, kids crowded around him and begged him to tell them stories about his three decades in the cab. They wanted to know how many horsepower his engine had and how fast it would go. They wanted to hear about his close calls and learn how every facet of his powerful engine worked. And after visiting with Kennedy, they always dreamed about growing up and taking a throttle in their own hands, just like their hero.

On this early morning of July 9, not very many kids gathered at the station when Uncle Dave climbed on board the Number 4 train. His huge black locomotive, steam already belching from the stack, would soon be headed west to Memphis. Rolling along behind it would be six cars filled with passengers, as well as two other cars carrying freight and mail. By today's standards it was not a big train, but in 1918 the Number 4 represented about as fine a ride as a person could get out of Nashville, Tennessee. At least the scores of frightened men leaving home for the first time would be riding in style.

While America's involvement in World War I had created some tough times for soldiers and their families, it had been a boon to the railroad industry. Troops were universally shipped to boot camps, stateside duty posts, and overseas debarkation centers via rails. Besides the hundreds of thousands of men in uniform, millions of others used the rails as a means of getting to new jobs in the war plants that had sprung up all across the nation. With scores of trains sold out many days before their scheduled runs, a rail ticket was a cherished item. Such was the case with the Number 4. Every seat was taken and a host of passengers would have to stand in the aisles.

As the minutes ticked by, Kennedy continued to watch nervously as more and more passengers crowded onto his train. A stickler for promptness, he wondered why they hadn't gotten on earlier. Shaking his head, he stared as fireman Kelley went about his duties. They both knew that with the full load they were going to need a lot of steam to bring the train up to speed and make up the time they had already lost. Uncle Dave also knew that this was not going be an easy run. It never was when a train started late.

Finally, at 7:07, the engineer got the go-ahead signal from conductor J. P. Eubank. Releasing the break and engaging the motor, Kennedy allowed himself a brief smile as the big engine lurched forward. Behind the shiny black locomotive, moving ever so stubbornly at first, the remainder of the train clanked and screeched down the tracks.

"All aboard," Eubank yelled as he stepped up onto the now moving platform. Finally, the conductor thought, we are on our way.

J. P. Eubank had worn his blue uniform for thirty-six years. He loved riding the rails; it was as much a part of him as breathing. Appropriately, his nickname was Shorty. Eubank had to stand on a bench just to pull the emergency cord. Yet even though he was small in stature, with his firm voice, snow-white hair and blue cap, there was no doubt that he had control of any train he stepped on to. No one, no matter how large, challenged this conductor.

The first cars Eubank would be working were filled to standing room only. The men and a few women who had been crowded into the old wooden coaches were all African-American. In 1918, blacks were still considered second-class citizens in most of the country, and therefore, no matter their station in life, they were not allowed to ride in the finer metal cars near the back of the train. Thus, seated on uncomfortable wooden benches, with steam and soot often blowing in through open windows, filling the car and making it difficult to breathe, scores suffered through—more than enjoyed—train travel.

The Number 4 was barely moving when Eubank had pushed all the way through the crowded aisles of the first car, looking at each ticket and then punching holes at the proper points on the cards. As he efficiently worked, he glanced out into the rail yards. The Union Station dispatcher had given the three-decade rail veteran something more than his passengers to worry about. The night train from Memphis had not yet arrived, and so the conductor was supposed to watch for it. Yet with so many passengers demanding his attention, and being required by railroad rules to have all the tickets punched by the train's first stop less than half an hour up the tracks, Eubank barely had time to look up.

While the engineer had the glamour job, the railroad companies considered the conductor the most important person on each train. He was the man in charge. If something went wrong, it was his responsibility. Though he wasn't "driving" the train, he was in a very real sense the captain of the ship. Thus, Eubank had to know if and when the Number 1 passed. It was only when the Memphis train cleared the tracks that the conductor could be assured that those who had chosen this train would really be safe.

Calling the train's porter to his side, the conductor waved his hand toward the window and explained, "Number 1 is running late, so we need to keep an eye out for engine 281. Let me know when you see it pass us."

George Hall was the young black man who was told to look for 281. Like Eubank, the inexperienced porter had little time to search for the train from Memphis. Yet he promised the conductor that he would let him know when the Number 1 passed by.

As the conductor worked his way through the cars, he stopped from time to time to greet old friends. One, Josiah L. Shaffer, a member of the famed Civil War group Morgan's Raiders, rode the train often. He was seventy-eight but still in robust health. He used his ticket to get a few miles out of town to fish on the Harpeth River. On this day, sitting with him was William Knoch, the wire chief of the Western Union plant and the older man's son-in-law.

Milton Lowenstien, a salesman for Jonas and Company, was seated not far from Shaffer, as was Milton Frank, president of the National Bag Company. After the two men visited for a few minutes, they decided to get up and move to the smoker car. But when Frank got out on the platform and looked at the mass of humanity that was crowding around the bar, he said good-bye to his friend and returned to his seat. Undaunted by the wall-to-wall men in front of him, Lowenstien pushed through the crowded coach and found a seat up at the front.

One of those standing in the aisle was Willis M. Farris, a much-honored Nashville citizen who had made a fortune in the lumber industry. Many credited him with putting the city's lumber industry on the map. The businessman shifted from side to side as the train picked up speed. He was obviously having a tough time keeping his balance. While others ignored Farris, a young man recognized him and sympathized with his situation. He signaled to Farris and offered him his seat. As the older man sat down, the younger one made his way to another car toward the rear of the train.

As Eubank continued to work his way through the passenger compartments, Kennedy and Kelley were building up speed. The Number 4 was moving at twenty miles an hour when it moved around Packing House curve. Looking up, Uncle Dave waved to people in cars and buggies as they crossed a bridge in front of him. Most waved back.

"Can't build up too much speed until we clear the yards and meet Number 1," the engineer reminded his fireman.

Because the Number 1 was already half an hour past due, both men in locomotive 282 expected to see the train very soon. They figured they would meet the Number 1 between the station and the Charlotte Avenue shops. If they didn't, as much as they would hate to, they'd wait at Charlotte Avenue until the Number 1 had passed.

As Kennedy checked the gauges he heard the whistle blast of an oncoming train. Glancing out the cab, he noted a locomotive pulling ten cars. As they met, Uncle Dave waved at the other engineer, then quickly noted the passing locomotive's number—381. Pulling out the morning schedule from his pocket, he glanced over to the listing for the Number 1. Uncertainty clouded his expression.

"Was that 281 or 381 that we just met?" he must have wondered as he studied the schedule. Kelley would be no help; he had been too busy stoking the fire to look up. Even more than eight decades later, no one is sure what the engineer decided. If Uncle Dave did misread the number of the switch locomotive and had not noticed that all its cars were empty, if he had thought that this was the overdue train from Memphis, then the engineer would have believed the track in front of him was clear. If that is what happened, the misreading of one digit had set in motion the events that led to the most disastrous train wreck in U.S. history.

But, as some assumed at the time, if the crafty veteran knew the schedule and correctly read the train's number, then over the next few minutes Kennedy undertook a series of actions that could rightly be considered criminal acts. Many feel that because he was running late, Uncle Dave might have tried to race ten miles to a second section of rails in order to make up the time. To this day, no one really knows which scenario put in motion the fatal crash between the Number 4 and Number 1.

As the Number 4 approached the yard control tower, Kennedy blew his whistle to let the tower operator know that he needed a signal. J. S. Johnson, who had been on duty for only eight minutes, looked up at the massive locomotive, waved and then hit the switch that dropped the green sign. Everything was go and Uncle Dave steamed on by the wooden tower.

Johnson picked up a pencil to record the event. As he wrote "No. 4 passed tower 7:15 a.m.," his hand suddenly froze. Searching through the night operator's log notes, he could find no entry that the Number 1 had passed. In a panic he shifted the signal to red, which ordered the Number 4 to stop, but the locomotive was already past the tower, so Kennedy could not have seen the signal change. Nevertheless, Johnson knew that either the conductor or porter would note the red board and stop the train. As the train continued down the tracks, the tower operator hit the key of his telegraph machine to contact Union Station dispatcher C. D. Phillips. Johnson quickly tapped out a message and then anxiously waited for a reply.

A few seconds later Phillips wired back, "Number 4 supposed to meet Number 1 there. Can you stop Number 4?"

In horror, Johnson reached for the emergency whistle. Pulling the cord, he set forth a blast that could have awakened the dead. But with the train now almost a half mile beyond the tower, no one heard the warning.

As he watched the train quickly drift out of sight, Johnson wondered, "Who is going to stop them now?"

In truth, if he had not seen the Number 1 for himself, if he had realized that the train they met was a switch train, then Kennedy should have stopped. The green signal was only a sign that the track was clear in the yard, not beyond. Therefore, the decision to continue was the engineer's. The conductor, who was still punching tickets, was the only one who could have overruled him. Eubank later admitted that he didn't see a train pass. Neither did his porter. They were simply too busy trying to take care of their passengers.

More than thirty miles to the west, sixty-three-year-old William F. Floyd, an engineer on 281, was more distressed than usual. This was his last run; he was retiring that day. But the extra passenger load created by the war had also made his Number 1 train from Memphis to Nashville late. As fireman Luther Meadows stoked the fire, Floyd considered the manner in which his railroad career was finishing. He was not just going to be late, he was going to be more than half an hour late. In his mind this was disgraceful. So every minute he could cut off the run made him feel a little better. Constantly checking his watch, he opened up the throttle in order to push his train beyond a mile a minute. Still, even as the central Tennessee landscape flew by at sixty-five miles an hour, the old man grew even more antsy. He wanted more speed. Meadows and baggage master Tom Dickinson could see just how upset Floyd had become and they felt sorry for him. They knew that no engineer wanted to be late on the last run of his career.

Floyd's train consisted of five day coaches, two Pullmans and a baggage car. As the engineer pushed his locomotive to the limit, the cars were literally swaying along as they approached Vaughn's Gap. The two front cars carried more than one hundred African-Americans who had been recruited in Memphis and Little Rock to work for Mason & Hanger on construction jobs at Du Pont's Old Hickory Powder Plant. They were covered with ashes and soot and couldn't wait to get off the train and breathe fresh air. The remaining passengers were mostly military men on their way to their duty posts, with a few businessmen, salesmen and railroad employees sprinkled in among the doughboys.

Ben Tucker, a white-haired, forty-seven-year railroad veteran, was one of those who had gotten a free pass back to Nashville. As he looked out the window, J. D. Thompson, an off-duty brakeman who was riding next to Tucker in the third coach, couldn't believe the speed that Floyd was getting out of the engine. As the countryside flew by in a blur, Thompson observed, "That boy is really highballing."

Even though they were going faster than usual, no one complained. After all, even to the railroad men everything seemed normal as the Number 1 passed Belle Meade and made the final stretch to Union Station. Many on board figured the next sounds they heard would be the squeal of brakes and the conductor's voice shouting, "Final stop, Nashville."

At 7:19, as the train hit a long bend, passengers on one side looked at White Bridge road. A minute later they noted St. Mary's Orphanage. As the sun drenched the Tennessee hills in a warm yellow glow, it appeared that it was going to be a beautiful day.

In the cab, looking out the window, William Floyd squinted into a sun that was still low on the horizon. As the engineer looked ahead, he observed a clear track to the point where it curved out of sight. That curve, now known as Dutchman's Curve, then called Dutchman's Bend, was just five miles from the heart of Nashville and surrounded by green cornfields.

At about that time, Robert D. Corbitt, the brakeman for the eastbound Number 1, inexplicably decided to leave the engine and check out the rear of the train.

As Floyd and his locomotive snaked around the curve, blind to what was ahead, Kennedy and engine 282 were hitting the beginning of the curve less than a half mile to the east. It was just twelve minutes since the Number 4 had pulled out of Union Station. Each trying to make up for lost time, the trains were chugging along at more than sixty miles per hour.

Kennedy probably saw the Number 1 before Floyd had a chance

to see the Number 4. Those who investigated the accident felt that Uncle Dave hit the brakes and killed the throttle, while Floyd probably did neither. Yet a second's delay on Floyd's part really made no difference. What had been set in motion when the Number 4 passed the tower could not have been stopped. There were two trains on one track and it was simply too late to change that fact. Fortunately, few of the passengers realized what was about to happen.

The eighty-ton Baldwin steam engines were still rolling at a mile a minute when their cow catchers came together. Like two elks in the forest, the locomotives went nose to nose and instantly went from sixty miles an hour to a dead stop. In a split second, benches were ripped from the floor, passengers were thrown into each other and sandwiched together with so much force that rescuers later found some bodies that appeared to be fused. Windows blew out, baggage became airborne missiles, and heads hit together in sickening thuds. On both trains, passengers in the first car looked up and found themselves literally pushed through the second and third cars. In a sense, both trains were folding up like telescopes. Then the "bomb" went off and scattered the already destroyed trains over a square mile.

The explosion caused by the two boilers simultaneously rupturing could be heard two miles away. As the trains were hurled through the air, wooden cars crumbled and lunged sideways. Some exploded, while a few of the back coaches hung precariously over the embankment. As if hit by an earthquake, the ground shook and the waters of nearby placid Richland Creek whitecapped against the bank.

Bodies and body parts flew out of passenger cars like confetti and littered the sides of the track. Steam floated over the scene like a thick fog. Through the sound of the hissing steam, those who had survived the colossal collision heard terrifying screams for help and moans of agony. Farmers who had been working in the nearby field raced over to the wreck and began reaching for victims. Many of those they first tried to lift came apart as they were picked up.

One farmer told the *Nashville Banner,* "You couldn't tell one part of the bodies from another. They were just all cut to pieces."

Engine 281 and 282 fell on either side of the track, unrecognizable masses of twisted iron and steel. The ruptured boilers were found well away from the locomotives' frames. The express car of the eastbound train drove through the flimsy wooden coaches loaded with human freight like a warm knife through butter. Those cars shattered as they rammed into the smoking car. The passengers in those first coaches had little or no chance at all. Most died; none escaped without injuries.

Because of the flying metal, wood and glass, blood ran everywhere. In an instant passengers had lost heads and limbs. Two-by-fours were driven through men's chests and scalding water splashed on hundreds when the boilers exploded. It took more than an hour for medical personnel to begin arriving, and even then there was very little they could do.

The smell of charred cars and the sight of bleeding bodies lying in the morning sun clad in uniforms evoked powerful battlefield images. The cornfields on both sides of the tracks were littered by fragments of iron and wood as well as body parts. Mail floated through the air like snow. People were pinned under cars, grotesquely sprawled across the ground, submerged in the creek and even hanging in the trees. Most of the dead were the blacks who had been the passengers riding in the first two cars on each train.

As soon as Union Station was notified, word was sent out for help. Within an hour hundreds of volunteers had come to help sort out the living from the dead. As they worked in the carnage, bodies were set to one side of the tracks and stacked beside the heaps of iron and wood.

As more of the debris was removed, workers discovered that many people were trapped under the cars, pinned to the rails. Though they tried, without heavy equipment there were simply not enough men to lift the wreckage and ease their suffering.

By noon on July 9, fifty thousand spectators had turned out to hear the moans of the dying and watch horse-drawn wagons haul away the dead. Roadside stands were selling food and drinks, and children were gathering up souvenirs. It was a grisly scene, made even worse by the thousands who were treating it as a kind of moving-picture show.

It would take more than a day before all the living had been found and freed. It would take three days for all the dead to be taken away. By July 12, body parts were treated just like pieces of the wreck. Railroad workers simply tossed severed limbs into big tubs and buried them beside the tracks. Getting the tracks rebuilt and the trains moving seemed far more important to the officials at the Chattanooga and St. Louis Railway than did finding what arm, leg or head went with bodies that had already been hauled off.

To handle the dead, embalmers were brought in from surrounding towns and coffins were shipped in from other states.

One of those taken away in a wagon filled with bodies was Robert Corbitt, the brakeman who had left engine 282 just moments before the crash. As an undertaker went to work on his body, Corbitt suddenly moved. Immediately taken to a hospital already overcrowded with injured, the brakemen recovered and continued to work for the railroad until retirement. He was one of the few lucky ones.

Even though doctors came from all over Nashville and other communities, there was little they could do at the wreck. About the only people who faired well were the ones in the metal Pullman cars, and even those men didn't all live. Willis M. Farris died, though the young man who gave the legendary businessman a seat lived. Josiah L. Shaffer may have survived the Civil War battlefields, but not Dutchman's Bend. Milton Lowenstien made the fatal mistake of going to the smoking car, while his friend Milton Frank's choice to avoid the crowd saved his life. Scores of others, black and white, young and old, became names on the death lists. All four men in the two locomotives died instantly. Tellingly, Uncle Dave Kennedy was found with a train schedule in his hand. To many, this appeared to indicate that he had been trying to outrun the Number 1 to the next double track at Harding Station. If that was so, then the engineer's gamble opened a large door through which the grim reaper gladly entered.

The final death tolls are still disputed. Officially, the Interstate Commerce Commission, the investigative body for railroad accidents at that time, listed the dead at 101. The railroad would tally the number of dead at 89, while the newspapers gave the final total at 104. Yet many witnesses believed those killed numbered twice that many. If so, why were so many dead not reported? Probably for two reasons.

The first was to play down the already incredible tragedy. Americans felt safe on the rails. Railroads could not afford having those who bought tickets suddenly questioning or losing faith in this mode of transportation. With automobiles already cutting into rail traffic, any negative publicity was harmful to the entire industry.

Secondly, the fact that at least eighty percent of the victims were black meant that many Americans simply didn't care as deeply as they would have if all the victims had been white. In 1918, African-American lives were not considered as being as important as the lives of white Americans. So identifying their bodies and having a full count of black victims was not considered as important as cleaning up the debris, fixing the rails and putting the trains back on the tracks.

Within a few days an official investigation was launched. One of the first witnesses called was Conductor J. P. Eubank. Asked why Kennedy didn't wait for the Number 1 to pass, Eubank replied, "He didn't think that there was anything to wait for." Did that mean that Kennedy had thought the other train had passed or that he could beat it to Harding Station? Eubank had no idea. Though clearly not at fault, the conductor would lose his job. Still, even with the firing of Eubank, no real blame was ever fixed on anyone.

Within three days, with many victims still not buried, the worst rail disaster in United States history disappeared from newspapers in Nashville and across the country. Some speculated that World War I was too dominant a story for much of the nation to bother with a train wreck. Historians now think that racism and the fact that most victims had been African-American may have been a factor as well. It is more likely that the railroad could not only generate steam in its locomotives, but heat on the media, politicians and law enforcement agencies as well. It was this influence that probably ended further coverage.

Few survivors remain alive today, few who remember firsthand the day when the Number 1 and Number 4 met on the tracks at Dutchman's Bend. But in Nashville there are still many who can't look at a railroad track without thinking of that day and what happened. The accident will probably never be completely forgotten thanks to a song penned by legendary Music City songwriter Bobby Braddock. The man who wrote one of the greatest songs in country music history, "He Stopped Loving Her Today," entitled his ode "The Great Nashville Train Wreck." It never became a hit, but those who have heard it are haunted by the imagery that hovers over each phrase and note, as Braddock's tribute to a black day in history continues to keep the memories of this senseless tragedy alive.

July 9, 1918, a morning like so many others until 7:19. What happened at that moment caused many Americans to no longer feel completely safe as they boarded a train to Memphis, Nashville or anywhere else.

# 5

# Killer Camille 1969

On August 15, 1969, Woodstock was finally winding down. The music fest had seen hundreds injured and two die taking part in a muddy celebration of sex, drugs and rock and roll. The news media had loved the debauchery and covered it as if it were the most important story of the day rather than just an outdoor rock fest. Ironically, Woodstock would stay in the public's mind much longer than the other two big stories that happened that same day.

In South Africa, Dr. Philip Blaisberg died. The fact that this man had been alive at all was the real story here. He had survived for nineteen months with another person's heart. No transplant patient had ever lasted as long. News stories reported that several of his days had even been close to normal. It seemed like a miracle to those who a couple of years before couldn't have even conceived of this kind of procedure.

Another story was printed that day, buried on the back pages of most papers but not even mentioned on national newscasts. Three people had been killed on the western coast of Cuba as a hurricane named Camille swept past the island. Beyond the tragic deaths of those three, the storm did very little damage to the island nation that was the focal point of Cold War America's darkest fears. Millions had felt that Cuba would be a source of death and destruction like the United States had never seen. On this day the government and its people would have been better served to pay attention to the storm that skirted Cuba rather than to Castro's threats.

Camille was the third hurricane of the season, and as it developed it appeared to be nothing more than a mild nuisance. The National Weather Bureau had even called it a feeble storm, little more than a tropical wave. In the forty-eight hours before it struck the Mississippi coast, Camille did nothing but meander around in a north-northwestly direction. Many forecasters laughed at the indecisive nature of the storm and pointed to its tepid winds and the smattering of rain it produced as a sign of its impotence. In most minds, this storm didn't even merit a name. Although it would still have to be watched, few figured it would amount to anything more than a footnote in storm history.

As it drifted over the Gulf's warm waters, Camille got a shot of energy, an unexplained injection of power, and that power drove it not west, as had been predicted, but north. Though few were noticing, the hurricane was beginning to take on the mantle of a killer and it was meandering no longer. As it slid over the Gulf of Mexico, its personality silently changed—its rotation tightened, its winds grew stronger, and it began to howl. In a matter of hours, the mild-mannered disturbance grew angry and powerful. Within twelve hours, it would go from being dismissed as a weak sister to a tropical storm to one with the potential to challenge Hurricanes Audrey and Betsy for destruction and mayhem. Audrey was a vicious storm that rampaged through Texas killing almost four hundred people and causing a mass evacuation of most of the southern part of the state. Betsy was not a mass killer, taking only seventy-five lives, but as she roared across a well-prepared southern Florida in 1965, the category-three storm caused massive destruction and flooding. Ultimately, even those two would not measure up to Camille. The only American storm that could rival it would be the Galveston hurricane of 1900. And Camille, like the Texas storm, would disguise itself as a lamb but become a ravenous lion. Before it had run its course, millions would feel its bite.

In an era when even Americans in the central plains follow the paths of hurricanes over the Weather Channel, it is hard to fathom that in 1969 most still "watched" the progress of storms via the newspapers. When the morning papers wrote of the weak storm that was casually aiming at the Texas coast, few in Louisiana, Mississippi or Alabama found any reason to be concerned. Even residents of the Lone Star State were not worried. According to the reports they read, if Camille did hold together until it made land, it was far too mild to create anything more than some light flooding. Yet the stories in the newspapers were now completely out-of-date. In the hours since they had been written, Jekyll had become Hyde.

Saturday was just another day for most residents along the Gulf Coast. Stores were open, people were shopping, the beaches were crowding with tourists and life proceeded as usual. A few newcomers and tourists checked in with locals and asked about what to do if the storm veered away from Texas and moved toward them. One hardware store owner in Biloxi spoke for most when he said, "Best way to prepare for the storm is to get yourself a bottle of good whiskey and sit back and relax." And many took that advice. Yet as the national forecasters upgraded Camille and some began to predict a turn to the north, a few looked out at the Gulf and considered all the options.

Radio stations began to issue warnings that Camille might be building into a storm to be reckoned with and that people along the beach should move inland. Television stations broke into Saturday-morning cartoons and issued similar warnings. Yet it just didn't look like hurricane weather to many locals. Besides, no storm could change that drastically in the time Camille had before it hit the coast.

By evening some fifty thousand Mississippians had taken the changing forecasts seriously enough to decide that the storm had the potential to be dangerous. Though many neighbors made fun of them, these folks boarded up their houses, loaded up family and pets and headed inland. To them, it seemed like a good time to vacation in Jackson and Hattiesburg, but even they couldn't have expected what would hit their homes in just a few hours.

As those with no stomach for high tides and rain fought bumper-to-bumper traffic on the highways, others prepared to greet the evening in a much different fashion. Some tempted fate by going fishing; others invited friends for parties, and some even hit the beach for a swim. In Louisiana's Lake Pontchartrain, four friends enjoyed the light afternoon breezes as they floated across the open water. They thought the worst that could happen would be a thrill ride across some fairly angry waves. The charcoal skies that hovered overhead just didn't look that deadly to them. Millions of others felt the same way.

A good portion of the people in small towns like Buras, Louisiana, population three thousand, and Waveland, Mississippi, population eleven hundred, also decided to stay put. They had been through storms in the past and they thought they knew how to judge them. While they respected hurricanes with real power, to them Camille seemed nonthreatening. Most expected to lose their power and phone service for a while, maybe have some minor damage, but not anything else. As they waved good-bye to the ones who decided to move inland, their faces showed little concern. Many even seemed to have a twinkle in their eyes, almost as if they were looking forward to seeing what the storm could dish out and how much they could take.

In Buras, the stores remained open, catering to those who had decided to ride this one out and welcoming thirteen unlucky newcomers. This team of men, some who had never even been to the coast, found out that their rides to an offshore rig had been canceled as a storm precaution. They were stuck and unhappy. With no other options open to them, they had been forced to get rooms in a Buras hotel. Locals told them that they had arrived just in time to experience a real treat. The men weren't too worried about the coming storm, they were just not happy about being stranded in a place with nothing to do. In their minds, all that lay ahead of them was a boring night.

Similar scenes were played out in other waterfront communities. Some folks left, some stayed. Some vacationers even thought that this would be a thrill, something to brag about back home, so they bought flashlights and film and got ready for the spectacle. Many honeymooners couldn't think of a better way to initiate a union and stayed in their beachfront motels. Thousands along a hundred-mile stretch of beach gladly traded common sense for a macho attitude and placed themselves directly in harm's way. Yet of all the people who seemed to shake a fist in the face of fate, the citizens of Pass Christian, Mississippi, seemed most bold.

In Pass Christian, a partylike atmosphere erupted almost from the instant it was learned that Camille had taken aim at the coast. It was almost as if the circus was coming to town. For the children of these families, it would be something they could talk about at school. The storm seemed so innocuous that at a beachfront Richelieu apartment, a luxury complex that in the past had repelled many hurricanes with its brick walls, a group of friends decided to throw a party in Camille's honor. They planned to greet the hurricane with drinks and snacks. Before she was finished, Camille would crash that party and many others, as well as crashing down on hundreds of miles of American landscape.

That evening, as the storm neared, experts contacted the radio stations with new information. It appeared that the storm was going to move west of Pass Christian. It now seemed that the community of six thousand was going to miss all the fun. The news darkened the spirits of those who had gathered for the beachfront hurricane parties. Looking out through the light rain, they figured that they had stocked up on booze and chips for nothing. The more than two thousand residents who had left town and the almost five hundred who had gathered in the high school gym were beginning to feel foolish too. At the Trinity Episcopal Church, where a few had gathered seeking shelter from the storm, folks came out into the mist and began to share stories of past hurricanes. A few even packed up their cars and headed for home. It seemed like Camille really was going to be a dud.

Still, local officials weren't nearly as confident as most residents. Although the storm appeared to have its sights set on Louisiana,

Pass Christian police cars headed up and down the coastal strip trying to convince those who were waiting near the beach to move to higher ground. When the unworried locals pointed to the Gulf, where waves were barely lapping at the seawall, officers shrugged shoulders and gave up. On the second-floor balcony of the Richelieu Apartments, a man with a drink waved and invited the cops to come back and join the party when their shift ended. The policemen declined.

A hour or so later, Police Chief Jerry Peralta drove up to the Richelieu. Getting out of his car and staring into the now steady rain, he knocked on a door. He was greeted by more than twenty partiers, many already appearing intoxicated. When Peralta begged the folks to come to one of the shelters, they laughed at him. Looking at the children playing in the room, he strongly suggested that at least they should go to higher ground. The parents would have none of it, however. This was a family affair and they were not going to be cheated out of their fun. The only way they were leaving was in handcuffs. The chief knew that he couldn't order them to leave, so he asked if he could take down their names and the names of their next of kin. When he had written down that information, he wished those gathered luck, and walked out the door. He had other stragglers he wanted to visit before he went back to the station.

As he walked back into the night, Peralta was suddenly greeted by winds that had jumped from a steady twenty to more than fifty miles an hour. Shaking his head, he wondered how many more idiots were staying at the beach and if any of them would listen to his warnings. Didn't people realize what winds of a hundred miles an hour and waves driven by those winds could do to beachfront property? As he drove away, the images of the children at the party haunted him. These images would continue to haunt him for the rest of the night.

At 7:15, the twenty-one-mile-long Mississippi Bridge at New Orleans was closed. The winds were now too strong and waves too high to make travel safe. Lake Pontchartrain, which had been fairly calm a few hours before, was now in an uproar. With their boat being battered on all sides, the four men who had decided to ride out Camille on the lake were now rethinking their decision. It was too late. Within minutes their boat began to break up. The canvas over their heads shredded like tissue and in no time the winds picked up to more than a hundred miles an hour. With little warning, their ride simply fell apart. Holding on to broken pieces of lumber and each other, the quartet stared into the teeth of a killer, each believing that this would be the last night they ever experienced. The men would somehow survive, but many others would not.

Camille's forward surge skirted the mouth of the Mississippi River ninety miles southeast of New Orleans before moving into Chandeleur Sound at fifteen miles an hour. From the air it might have looked small and rather harmless, but the hurricane was in reality severely compressed, tightly wound and ready for action. For days it had been fooling everyone, including the forecasters, and now it was ready to show its true nature. It was tearing away its mild-mannered guise.

Without any real warning, some fifty miles to the southwest, the area from Buras to Venice was buried by the waves that jumped the fifteen-foot-high Mississippi River levee. It had been raining for a while and the winds had picked up, but since locals had battened down the hatches and made the usual preparations they felt safe in their homes. When the winds picked up to a hundred miles an hour, some began to wonder if they had made the right decision. They looked again to the morning paper, reread the story and studied the path where the storm was predicted to hit. The weather folks had misread this one, they realized as the winds howled and pieces of their roofs flew into the skies. For these poor people, it was too late to leave the peninsula, but they had another option. Before the storm rained down any more terror, many ran to the local school. It was a solid structure, built to withstand even the strongest storm, and had a second story. If the water got above the levee, at least there would be something between them and the surf. In the next few minutes, scores of locals fought through the rain and winds to wait it out there.

At the Big Fish Hotel, the stranded oil rig workers seemed a bit more worried than the residents. They paced in their rooms as the winds picked up. They didn't know about the school and wouldn't have known how to get there even if they had the information. They had never heard such wind, and the rain was falling so hard they couldn't see out their windows. "How much worse can it get?" someone asked. They weren't sure how much more the building could take.

A door blew in with a loud crash that scared them all. Then water started sweeping through the streets and into the hotel. As the men watched, it quickly cover the floor and they opted to move to the second floor. For the next fifteen minutes they peered out windows and down the stairwell as the Gulf came closer and closer to them. With the winds now howling at more than a hundred miles an hour, the thirteen sat alone in the darkness. Camille was shaking the Big Fish Hotel with each crashing wave. Windows were blowing out and the water was creeping closer to the visitors' perch. The dark, nasty water seemed to reflect the fear gripping the men. As it lapped toward them, it seemed to laugh in their faces.

When the water reached the second floor, the men crawled on top of the beds. Soon, as the Gulf pushed in through the upper-story windows, they realized that they were not high enough. If they stayed where they were, they would either drown or the beds would float out into the horror that was just outside the hotel. Though it was all that was protecting them from the mighty winds and torrents of rain, they hacked through the ceiling into the leaky attic. Pulling themselves up to the highest reaches of the hotel, they prayed and blindly hoped that the old building would somehow stand up against Camille's force.

Norman Ballivero was praying too. Stuck at his Western Auto store, he grabbed a flashlight and life preserver and climbed onto his roof. The water, now more than fifteen feet above the street, was filled with debris. Small boats, pieces of houses and entire mobile homes were floating by him as he tried to hang on to his perch. As stacks of refuse piled up against his store and 150-mile-an-hour winds whipped his body, he wondered if he would live to see the next day. He had never experienced a storm like this one.

If Ballivero had been able to hear anything but the wind, he might have noted his own son's cries. Not far away, the younger member of the family was clinging to his roof like his life depended on it, and it did. He would stay there the rest of the night.

Camille had placed Buras and its people in a terrible position. The winds and waves were overriding the town; the only ones who had any kind of security were stuck on the second floors or roofs of buildings. In the dark night, the storm had tossed a two-hundred-foot barge carrying combustible solvents into a nearby power plant. Live wires were being tossed back and forth on the barrels of explosive materials. If one of the tanks ruptured as the ship rocked back and forth, a single spark could create an incredible explosive force that might even outdo that of the hurricane. For the moment it appeared that if Camille didn't wipe the town off the shore a ticking bomb might.

For the next two hours almost everything man-made between Buras and Venice disappeared. Acting like a giant bulldozer, Camille kept pushing whatever it caught along the coastline. After being punished for four hours, the oil rig workers would look out from the attic of the Big Fish Hotel into a night that was still black but no longer hostile. Norman Ballivero and his son would survive too, as would all of those who sought shelter in the school. But the community they called home was gone, swept somewhere into the Mississippi River marshland. Left behind were oil slicks, piles of debris and thousands of broken dreams. While only one person would die in the two towns, damage was so extensive it was as if the area had been returned to nature. Still trapped by the waters, exhausted from the stress and fight to hang on during the storm's height, those who stayed in the towns had only the energy to grimly look toward the northeast and Pass Christian. They hoped that the people there were prepared for the hit they were about to take. If communication lines had been up they would have urged the citizens at Pass Christian to get as far away from the beach as possible.

Across the Gulf in Mississippi, Camille was still a few minutes from greeting those who were stubbornly waiting for her. Many here still thought the storm was a bust. Winds were now howling at sixty and the Gulf waters were beginning to top the Pass Christian seawall, but even a good thunderstorm could do that. Was that all that this hurricane could muster?

No one there knew about the blows that Camille had just struck in Buras and Venice, nor did they realize that the towns of Waveland and Bay Saint Louis, Mississippi, were now being pounded.

Waveland and Bay Saint Louis had been largely evacuated. Residents realized that even a small hurricane could greatly punish beachfront towns. With lowland and water all around them, residents could find few places to run when a big wave hit. Yet those who gathered up their belongings and drove inland could not have guessed the extent of Camille's wrath. The storm blew in with 150-mile-an-hour winds, tornadoes, and fifteen-to-twenty-foot tides. Camille also carried with it a great deal of debris that it had picked up as it had traveled at slow speed across Chandeleur Sound. Now even stronger than it had been when it devastated the small Louisiana towns, in a matter of moments Camille would level both Waveland and Bay Saint Louis. Ninety-five percent of their structures would be flattened. Those that weren't would not last long, as hot electrical wires would spark fires across the wide areas of destruction. The Bay Saint Louis Bridge, spanning the wide bay and linking the community with Pass Christian, would be so jammed with lumber, trees, dead animals and other refuse that it would be days before it could be used. It was a testament to its engineering that the bridge managed to stand. There can be little doubt that if those to the east had seen the images in Waveland and Bay Saint Louis, witnessed the storm's fury firsthand in Louisiana and seen the fear in the eyes of storm vets who managed to survive, they might have taken pause to pray or run. Yet without this knowledge, the storm was still a party waiting to begin for those in Pass Christian and other oceanfront cities.

At 10:15, the power in Pass Christian suddenly went out. Now the night was black. Most took the outage in stride. They had expected it and were prepared with lanterns, candles and flashlights. As people looked through windows into the August night, scores of dim yellow lights twinkled like stars in the light rain. It was serene, almost romantic, but that was just an illusion. No one realized that this was the literal calm before the storm.

A few minutes later, drops of rain pelted the ground like sprays of machine-gun fire. The noise from the huge drops was so loud that only a few noted the wind shift. Fear had not yet set in, but folks were finally beginning to get excited. As they looked out into the blinding rain, those at the apartment complex grinned and shouted over the roar, "The party is about to start!" Camille would soon bring an end to the carefree attitudes.

In a single instant, the winds jumped from 60 miles an hour to more than 180. The barometric pressure dropped so quickly that people covered their ears and screamed in pain. Their heads felt as if they were going to explode when the barometers bottomed at 26.61 inches. The wind, battering everything in its path, picked up signs, cars and even buildings, smashing them against each other. Fish, rocks and even boards were effortlessly tossed through the broken panes and doorways. In the darkness, those who had waited to greet the storm were initially shocked, then terrified. They couldn't hear themselves think and the rain was so heavy they couldn't see anything. Windows, doors, walls and even the roofs over their heads were being smashed as if they were toys. The world was coming apart at the seams. There was no place to run. No structure near the beach was safe. No man, woman or child who had stayed to experience Camille's terror now wanted to be anywhere near the storm. This was no party. This was a war—and one side had all the firepower.

In the next few minutes Camille would spawn as many as fifty different tornadoes, all spinning through the city at one time. Inside their funnels, winds of more than 300 miles an hour carved through areas and left nothing in their wake. Behind the small twisters came steady 210-mile-an-hour winds. And though no one would have believed it, the worst was still offshore.

In front of its eye, Camille was now driving a storm surge or tidal wave that was more than twenty feet high. It was moving at an accelerated rate and gaining steam with each passing minute. This was the storm's most lethal weapon, a nightmare of water that had the power of hundreds of bombs. It was aimed at the very center of the coast along Pass Christian but was large enough to deal death blows for miles on either side.

The Richelieu was right in the middle of the path. The men, women and children who had been so smug an hour before now felt the very foundation of their building move back and forth in the wind. Rain poured in through broken windows and collapsed doors. The wind was pushing grown men up against walls and pinning them there. Children were flying through the air like kites. No one could help anyone, and in the darkness no one could see what was coming. They were cursing and praying and crying too, trying to hold on to something that would anchor them for another minute. They figured that if they could just live for sixty more seconds, the worst would be behind them.

For a moment the winds and rain lessened, so Camille's knockout blow, the one that had targeted everything along the beach, didn't appear as frightening as the storm's initial wave of attack. Yet its rolling tidal wave—now almost forty feet high—was a monster, probably the largest the coast had ever seen. Contained within it were millions of tons of water and masses of debris. Easily riding over the seawall, the wave quickly destroyed Beach Avenue and then pushed inland. Like a bulldozer, it flattened everything in its path, apartments, hotels, resorts and homes. It turned streets into rivers and what was left of houses into boats. Ships that had been tied up offshore were pushed through the business district, taking out buildings along their route. And those who had stayed to watch the storm were being washed along with the trash. In a matter of seconds, the famed beach was scraped clean. The Richelieu and its party had been erased, its revelers buried under tons of debris or swept downtown in a torrent of raging water. Few of the partiers probably lived more than a few seconds. Most were smashed to death before they had the opportunity to be pulled under and drown. There would be no stories for the grandchildren and no pictures to send relatives. For these people, the party was officially over. But the storm was just getting started.

After taking care of the beachfront property, Camille headed inland. The Trinity Episcopal Church, a rock of a structure where so many had gathered to talk about past storms and wait this one out, was ripped apart. Many of those who had chosen to take refuge there simply disappeared into the rushing water. Across town century-old trees were uprooted and became missiles, cars and trucks became not only weapons of destruction but coffins as well. People died in masses and they died alone. Camille was taking no prisoners, and its roar was so loud that the screams of those who were caught in its grip went unheard.

A fisherman who had been blown off his boat tied himself to a bridge to avoid being swept away by the storm surge. He survived Camille, but when the waters receded as quickly as they had risen, the rope that had just saved his life pulled tightly around his neck. He must have struggled for several minutes before the execution was complete and his body hung limp, swaying in the wind. Others died just as horribly. One minute they thought they were safe, the next, Camille hit them with another blow.

Camille would pelt Pass Christian for four long hours, finally leaving at 2:00 A.M. Those who had somehow survived were greeted by waist-deep water rushing through the streets and the screams of victims who were injured or in shock. Hundreds of victims were gathered in the trees and on the roofs of buildings that had not collapsed. Many were mangled, most were frightened, but they were lucky to be alive. Next to them, as well as buried under the debris, were others who were not so fortunate. The sky was pitch black and would stay that way until the sun came up almost five hours later.

In the dawn's light those who were left in Pass Christian could finally assess the damage. Bodies were everywhere. Several, as if part of a sick joke, had somehow come to rest in a cemetery. Some were in homes broken like eggshells; others were buried in mud. In the town of six thousand, almost all businesses and more than three-quarters of the homes had been destroyed. There was no water, power or edible food. Roads were impassable. Dead animals seemed to be everywhere. As one resident told a *Newsweek* reporter, "Storms have hit here plenty of times before and we just pass them off. But this one really fooled us." The worst was over at Pass Christian, but Camille was hardly finished with its reign of terror. Before turning inland, it would storm up the coast for a while, taking shots at everything in its path, its outlying winds, tornadoes and tidal waves wreaking havoc and bringing more death. Camille's lethal threefold attack hit Long Beach and then Gulfport and Biloxi. The damage there mirrored Buras, Venice and Pass Christian. The towns were flooded, the beaches scraped clean, and the wind blew so hard that cars and boats flew through the air for blocks. Ocean-bound ships ended up blocks inland—and everywhere people died horrible deaths. Each community had unique stories and suffered similar hardships, but all stood in awe of the storm that they had written off.

Even in faraway Mobile, Dauphin Island and Pensacola, beachfront homes flooded and power was knocked out. In Mississippi City, century-old mansions that had survived countless hurricanes were reduced to splinters. Along the coast, bridges were lifted from their foundations and tossed around like paper. Debris was everywhere. U.S. Highway 90 became a dumping ground for everything the storm picked up—baby beds, chairs, lawn mowers, cars, lumber, boats, new clothing with price tags still attached. By the time Camille moved inland, the highway appeared to be a sixty-mile-long yard sale gone bad. The storm also pulled the concrete from the road in chunks. It would take more than two years for the coast's most vital traffic artery to be rebuilt.

With roads and rail lines wiped out, no one could get anywhere. For almost a hundred miles in either direction, everyone along the Gulf was without water, food, medicine, gas and power. Most would remain isolated, hungry and wet for at least forty-eight hours. Even as the storm headed north and people came out of their shelters, horror still awaited those left behind.

At daylight, the citizens of Pascagoula began to take stock of the damage. They had not been hit by the eye, so the city had structures left to rebuild. But what greeted them as they looked out on their flooded lawns and streets was a scene too scary for even the *Twilight Zone.* Hundreds of thousands of water moccasins had been driven from the swamps and into the heart of the city. They were angry, hungry and omnipresent. And they wanted to find a dry spot as badly as did the people. One woman counted more than two hundred in her backyard. Pascagoula and other cities would battle the poisonous snakes for weeks.

Within twenty-four hours of the hit, the National Guard rolled in trying to pick up the pieces. They were met with humanity gone bad. Merchants were charging a dollar for a nickel loaf of bread. Water was going for five dollars a quart. Stores were being looted. There was hardly the gentle beach attitude that had been evident a few hours before the storm. A curfew was ordered, but there was no place for people to go. The only way to get off the streets was to climb into a soggy car. Many lived that way for days, dodging snakes, and then, waves of rats. They drank warm beer and soda and lived with the smell of death all around them. Not only did humans perish, but Camille killed millions of fish and scattered them in its wake. It was worse than a war, worse that anything anyone could have imagined.

More than two hundred thousand along the coast were homeless, another four hundred dead. A plane carrying relief supplies crashed while trying to bring in food and supplies. All four on board died. The only good news in the midst of the rubble was that a five-year-old boy had somehow floated out of the Richelieu apartment complex, using a mattress as a boat. A gift from Camille, the child had somehow survived the hurricane and the party. The twenty-three others who had been with him did not.

Hours after the storm left the coast, the National Weather Bureau assured those to the north and east that Camille was a burned-out storm. It had spent it power and fury on the coast. It supposedly didn't have any punch left. Yet once again the experts were wrong and this time no one was surprised.

Camille, now downgraded to a tropical storm, spent the next few days slicing through Mississippi and Tennessee, spinning off scores of tornadoes as it dropped rain across the south. As it passed just north of Bowling Green, Kentucky, it finally appeared to be harmless. Finally, even those who had felt its wrath firsthand felt safe and secure. Then, again without warning, Camille changed its personality and exploded, driving torrents of rain into West Virginia and the Virginia hills. The revitalized storm would drop almost a foot of moisture on the mountains and in the valleys before moving out to sea. In its wake, flash floods would rip through the valleys along the James River and Camille would kill more than sixty people.

As he looked at its week-long marathon of destruction, Dr. Robert Simpson, head of the National Hurricane Center in Miami, called Camille "the greatest storm of any that had ever affected this nation, by any yardstick you want to measure with." He also was thankful that it was finally gone. Still others wondered.

As Camille headed east in the Atlantic, people who hadn't given it a second thought a week before followed the storm with a fearful eye. They knew that it couldn't turn back, they believed that it was now heading out to sea to die. After taking more than four hundred lives and causing hundreds of millions of dollars in damage, it was finished, but this storm had fooled the experts so many times before that millions still watched it and prayed. And finally, as they looked on, Camille died. For the first time in a week, their prayers were answered—the killer killed no more. But in Pass Christian, Buras, Venice and up and down the Gulf Coast, the locals had a new respect for Mother Nature. No longer did they see an approaching hurricane as a reason to throw a party and no longer did they feel safe when any storm approached.

# 6

# The Burning of a Coconut Grove 1942

Yogi Berra once described a New York restaurant this way: "Nobody goes there anymore, it's too crowded." On the evening of November 28, 1942, the famous baseball sage could have very well been talking about Boston's popular Coconut Grove nightclub. With music, food, liquor and wall-to-wall people, the Coconut Grove was the heartbeat of Beantown on this cold fall evening. In the lingo of the day, the joint was jumping as more than a thousand people had somehow crowded into a multiroomed hot spot that had a supposed capacity of four hundred to drink, dance, eat and relive the "big game."

A few hours earlier, Fenway Park had been filled to capacity as fans of the Boston College Eagles watched their heavily favored squad take on the undermanned Crusaders from Holy Cross. The mayor, local celebrities and even representatives of the Sugar Bowl were on hand, the latter group ready to present to BC an invitation to their prestigious New Year's game. It should have been a walk in the park for the Eagles. Their fans were so sure of victory that a party had already been set up for that evening on the terrace of the Coconut Grove. The Crusaders didn't care much about the national polls or the esteemed visitors from New Orleans; they didn't feel like being the reason for a local victory party either. Fired up after reading about the strength of the Eagles for a week in the paper, the Holy Cross squad drove into town and waxed Boston College 55–12. Forty-two thousand fans were stunned as they left Fenway. They could not believe what they had seen. To millions in the city and surrounding area who had listened to the contest on radio, the game seemed like a bad dream too tragic to fully comprehend. Yet as the local sports partisans would soon find out, that loss was nothing compared to the nightmare that would take place just a few hours later when fans from the two schools and several hundred other folks crowded into the city's best-known club.

One of those who really enjoyed the big game was Joseph Boratyn. A year before when BC and Holy Cross had met, he had worn HC purple. Now a Crusader alum, the former star fullback had gone to Fenway praying for a miracle. His prayers were answered, and as the game went on he wished he could have been on the field. As he left Fenway that day, Boratyn and some of his buddies decided to whoop it up at the Grove. It would be a great way to rub the victory in the faces of the BC fans. So with Joseph leading the way, a large group of purple-clad partisans wandered down to Piedmont Street.

Billy Payne was headed to the club as well, but he had a much different reason for hitting the club on a Saturday night. He wasn't there to drink and party, he was there to work. Payne was one of the stars of the Grove's famed floor shows. Billy's tenor was almost as popular as the drinks served at the club's several bars. As he walked in the building's back door, the crooner wondered if Boston College's loss would put a damper on the Thanksgiving weekend. Like most entertainers, he dreaded working a crowd that was nursing a deep psychological wound. He hoped that the many BC fans would bounce back better than the team had.

Wilbur Sheffield of Newton was also headed to the Grove. He wanted to get there in time to catch Payne, the chorus girls and to socialize with some friends. But the General Electric Company engineer was running late. With the war machine running at top speed, power demands were at an all-time high. Sheffield knew no normal work days and few days off. Always on call, he hoped to get away soon enough to catch at least some of the action at the club.

Long before the Grove opened for business, Stanley Tomaszewski checked in with his boss. Though too young at sixteen to get into the club, he had still been hired to bus tables and do odd jobs. With so many young men in the military, club owner Barnett Welansky felt a need to break the rules in order to secure enough help to keep the club running. Welansky's desire to break this law would ultimately catch up with him. Yet Tomaszewski was thrilled to have the job, and with a long list of reservations in the book, he was hoping that the crowd would be in a tipping mood.

As the rain fell and the hidden sun set, a gray cold embraced Boston. That same evening one of the most popular men in America tried to beg off of his promise to go to the Coconut Grove. Buck Jones was tired and sick, a cold dragging down his energy and enthusiasm. The Vincennes, Indiana, native had been a well-known movie star since 1917. With hit westerns such as *Riders of the Purple Sage, Riders of Death Valley, Pony Express* and *Down Texas Way,* he had been a cowboy icon since the days of silent movies. Now a member of a trio of stars who headlined Monogram's *Rough Rider* series, the fifty-three-year-old Jones had been sent east on a patriotic mission to help sell war bonds. The bond rallies he attended would have been demanding enough for most Hollywood stars, but Buck also felt a need to visit with very ill children. With Christmas so near, he gave a great deal of time at each stop in the critical hospital wards. That is how he had spent his morning. Then, just before noon, Buck went to the Boston Garden for a rally of about twelve thousand of his "children," the kids who were the real fans of the western star. From there it was a trip to Fenway, where he was interviewed by the local press as he watched Holy Cross beat BC. By the time the football game ended, he was exhausted.

On any other occasion, Jones would have probably stayed in his hotel and rested. His fans were not old enough to get into a nightclub and the Grove hardly fit the clean-cut western image that Buck had embraced in his pictures. But just like employees Payne and Tomaszewski, on this night Jones was expected to be there. A party had been arranged in his honor. Tables had been reserved and food had been ordered. Besides, because of the war, a lot of the boys who had watched Buck in matinees in the thirties—his old fans—would be there in uniform.

Boston College's loss had no effect whatsoever on the young marines, navy and army men who had just received their last weekend pass before being shipped off to Europe. The famed Grove, with its music, decorations and girls, was the stop for hundreds of the young servicemen who were heading out on this cold night. The old saying "Eat, drink and be merry, for tomorrow you may die" was for them a possibility and not merely a cliché. They were planning on drinking, dancing and flirting until they were kicked out of the club.

Along with scores of football fans, employees, businessmen and a movie star, by nine that night hundreds had jumped into cars, on buses or taken cabs to get to the Grove. With the whole world at war, none of them knew what the future held, but they were certainly not concerned about their fate this evening.

In spite of the war, or maybe because of it, Boston's most fashionable club was hopping early on this night. People were pushing through the revolving door of the main entrance on Piedmont, checking their coats and hats and wandering through the huge venue. With loud music playing, drinks flowing and wall-to-wall people on the dance floor, at the Caricature Bar, in the New Cocktail Lounge, the dining room, up on the terrace and down in the basement at the Melody Lounge, it seemed that Boston College's loss had been long forgotten. Sure some, like the mayor, had canceled to stay home and mope, but many others had come to celebrate, even if their team had lost.

Gathered around Joseph Boratyn, Holy Cross faithful were reliving Bobby Sullivan's first touchdown, Johnny Bezeme's sixty-seven-yard scamper for a score and the intricacies of a defense that completely shut down BC coach Denny Myers's famed T formation. It was truly a Thanksgiving holiday for Boratyn and company, and the menu consisted of grilled BC Eagle.

Buck Jones's spirits had picked up as well. East Coast theater owners had arranged the testimonial dinner for him at the Coconut Grove. Those around the cowboy star seemed thrilled to meet him. For Jones, who just a few years before had thought that his career was over, the enthusiasm the theater owners had for him on this night gave him a renewed confidence in his own future. He felt that he was still a hero, a star, and with the top cowboy draws such as Gene Autry entering the service, Buck and his horse Silver could again find a place on the big screens across America. Things were going so well that the only reason Jones was looking forward to getting back to the hotel was to call his wife and tell her about the wonderful reception. As Buck laughed and signed autographs, as people lined up to meet the famed man, some, in a distant corner of the club, decided to pull a prank. That seemingly harmless joke would set in motion a series of events that would not only spell the end to Buck Jones's career and life, but would initiate a fire that horrified the entire nation.

The Coconut Grove was a stubby, one-and-a-half-story, block-long establishment. Claiming a Piedmont address, the rear faced Shawmut Street, one end ran along Broadway, and the final side rested against other businesses. Most of the club was deep, but the area that included the New Cocktail Lounge was not nearly as wide, thus creating a complex L shape. The club featured a Hawaiian motif, complete with fake palm trees, fruit, and walls covered with brightly colored cloth. As was to be expected, it seemed that coconuts hung in almost every nook and cranny. The center of the club's world was the huge dance floor with a large orchestra platform and rolling stage located at the back of the complex. The sign on the front promised an incredible experience of dancing and dining, and those that entered this creation of palm trees and tranquil blue skies must have thought that they really had found another Oz.

Even during the day, when it wasn't open, navigating through the Grove was difficult. Hallways were narrow, the stairs down to the Melody Lounge were especially hard to negotiate and the very dark building appeared as if it had just been thrown together without forethought. In a way it had been. Because the club had become popular so fast, the owners had bought out other businesses, knocked down walls and haphazardly expanded to accommodate the large crowds. While great expense and efforts were given to decorating the facility, little thought at all was spent planning for emergency evacuation. Most frightening, while the Coconut Grove had several exits, at least one on each street, none were marked and most patrons knew only about the revolving door on Piedmont. In addition, the building had no sprinkler system and only a few fire extinguishers, some hidden in order to not ruin the Polynesian atmosphere.

Upon inspections, fire marshals had limited the club to just over four hundred patrons, but capacity codes were rarely enforced. No one counted heads and there was no real way to turn people away. So by ten o'clock, more than a thousand people had pushed into the Coconut Grove with even more expected soon. The club's owner was ready for the throng and had waiters set up extra tables and chairs on the main dining room dance floor.

In the real world, palm trees and blue skies are not so highly flammable. This was not the case in Boston. The trees and other decorations were made of cloth, paper, bamboo, and the blue skies overhead were satin. The walls, which appeared to be leather, were really a cheap imitation fabric. The dim lighting hidden in the coconuts hanging on the trees created the illusion of twilight, and few patrons could see well enough to have their fantasies dashed. It might have been cold and wet outside, but inside the Grove it was always summer and the skies were always clear.

Most of the servicemen found the dance floor the big draw. Some would sit on the raised terrace at the far end of the room, spot women dining at the tables, then climb down to secure their partners. In the summer, the roof was often opened, making moonlight dancing possible. But on this evening the rain and the cold weather meant that no one would be seeing the gray Boston sky.

As was expected, Buck Jones was a big draw. Hollywood royalty rarely visited Boston, lured instead by the bright lights of New York

City. But with the cowboy star so willing to pose for pictures and sign autographs, the main dining room was the center of the tropical world on this November night. One of those admiring the movie star was John O'Neil. The young man had just gotten married and his entire wedding party had arrived at the Grove to celebrate. Originally, O'Neil and his bride planned on leaving for their honeymoon at ten, but when Jones arrived, the couple decided to meet the actor and catch the next floor show.

As Jones signed autographs upstairs on the terrace, there was another kind of excitement brewing in the basement. Besides the dance floor and dining room, the club's other hot spot was the Melody Lounge. Though just off the kitchen, the elaborate decorations hid any facet of the real world of cooking steaks and dirty dishes from the patrons. Like everything else in the club, the small bar section in the basement seemed like a part of a tropical paradise. Noisy and usually filled with the college crowd, those who wandered down to this part of the Grove usually drank a great deal more than they ate. Tonight Holy Cross football fans were toasting each touchdown time and time again. For most of the night it seemed harmless, but the mix of booze and victory soon became too much. As the group grew rowdier, they began to laugh and horse around in a much more destructive fashion. A few patrons smashed glasses and others treated the ladies a bit too roughly. One of the drunk students even climbed up on the bar and removed the one overhead light in that part of the club. He tossed it to a friend and they staged a game of hot potato in the near darkness. Within seconds the lightbulb had been smashed. With only the dim mood lights illuminating the room, the group became even more unruly. Sensing that things were getting out of control, the manager ordered Stanley Tomaszewski to climb up on the bar, then onto a chair to replace the light.

The teenager became the floor show as some patrons began to cheer for him and others made fun of the kid. In truth, many wanted the light to stay off, and some even threatened to knock Stanley off his unsteady perch if he didn't get down, but Tomaszewski did not shrink from his duty. Carefully balancing himself, he stood on the chair and felt his way along the low ceiling for the socket. Several times he tried to find it without success. Finally, as the noise grew louder, he reached into his pocket and pulled out a match. Striking it, the small flame lit up the boy's young face. Seeing the empty socket, Tomaszewki quickly twisted the new bulb into place. He then shook his match out.

As Stanley crawled down, someone noted a small dark hole on the ceiling. It seemed to be growing. Seeing a puff of smoke, the bartender grabbed a wet rag, climbed up on the bar and hit at the tiny flames. Those around him watched the man try to douse the flickering fire with a detached interest. Even though he was not making any progress and the hole in the ceiling seemed to be growing, to minds numbed by booze, the danger did not sink in. Sensing the bartender needed help, another employee found an extinguisher and aimed it at a palm tree that had now caught fire. Still there was no panic. Unable to reach the fire in the high branches of the tree, the man tried to pull the palm down on the floor. It took less than a minute to bring the tree down, but it was too late. The silk curtains that hung from the ceiling were now aglow. The fire really took off when a patron reached up and tried to pull a piece of flaming cloth from the ceiling. When he cried out because his hands had been burned, the seriousness of the situation hit some of the drunken revelers. Though they had been watching the fire for a couple of minutes, they finally realized that this was not part of the floor show. When the flames burned through an electrical wire and the room was plunged into darkness, a sudden and infectious panic spread throughout the room. Within seconds people were screaming and everyone was trying to make their way to the narrow winding stairs at the far end of the room or navigate the room's other stairwell that led to the street.

One second the fire had been barely alive, burning just a tiny part of the fabricated sky, a few seconds later it was raining sparks onto the heads of the fleeing customers. As it spread, a loud chorus of screamers cried out, "Fire!" Those who had just curiously watched the first flames now found themselves a part of a confused and panicked mob. With the Melody Lounge encased in a choking darkness, it was almost impossible to sense which way was out. Everyone at once was struggling to reach either of the two stairways. In the mad rush, many were knocked down or shoved aside. Friends who had just been slapping each other on the back were now literally climbing over each other trying to get away from the rapidly spreading flames. The employees had the advantage, as they knew the route to safe passage, but to get to that route they were having to literally fight the terrified crowd.

A few saw the hopelessness in trying to flee up the stairs and headed for other lesser-known exits. Some ran into the kitchen and escaped though a service door. Others waded through the smoke and found some of the other unmarked exits out through the basement. But most seemed intent on battling their way up the two sets of stairs.

With fear hanging as heavy as the smoke, fights broke out. Drunken men, hit by the sobering thoughts of burning to death, slugged anyone standing between them and their chosen escape route. As flames roared overhead and charged across walls, as bottles of liquor exploded behind the bar, punches were thrown from every angle. Friends fought friends for the right to live. Yet as the flash fire quickly consumed the room and a number of victims were quickly overcome by fumes, a few simply gave up. Unable to see through the smoke, realizing that they would never get through the wall-to-wall bodies that were jamming the stairs, these hopeless revelers simply returned to their chairs or stools, some even lifting a final glass in a toast to what might have been and waited. They didn't have to wait long—they perished without a struggle just three minutes after they had seen the first evidence of fire.

The stairway to the street was a grotesque place. Women were tearing at dresses that were now aflame. Men were trying to snuff out fire that had ignited in their hair. The struggle to reach the stairs seemed worth it, because if they could make it up those thirteen steps they would be safe. The first man to make it to the top hit the door and was knocked back. Confused, he hit it again. Soon others joined him. Unbelievably, the door was locked from the outside. Kicking and screaming, several men and women tried to push their way outside, but to no avail. As flames crawled up toward them, a few prayed, many cursed, but no one could open the door. There was no time to turn around, no time to realize that they had been locked in because so many had snuck down to the Melody Lounge using this back entrance in the past. Since there was a strict policy that no one was to enter or leave Boston's finest night spot without paying, the door had been locked and barred to turn away free-loaders. Tonight this policy proved deadly.

The frantic survivors who had chosen the stairway to the upstairs portion of the Grove were also racing the flames. As they emerged from the stairwell, they ran through the foyer past the cloak room and toward the main exit. There they faced the revolving door. An employee was trying to get the crowd to calm down and exit one at a time. It was no use. Those who had escaped the inferno in the basement were mad with fear. In a panic they all tried to be the first to rush out into the cool night air. A few made it, but when one fell and the doors stopped spinning, others tried to climb over him. This created a wedge that jammed the door. Soon screaming men and women were being pushed down one on top of the other. The wedge had grown six bodies high and no one could be convinced to stop and help to lift the others and exit in an orderly fashion.

Outside the chaos in the cold night air, Wilbur Sheffield had finally made it downtown. He parked his car and was heading for the club. As he walked down the street he noted a local fire brigade battling flames that were consuming a parked car. Sheffield watched them for a few moments, feeling sorry for the person who would soon leave the club and find his car a burned-out shell. As he turned and again approached the Coconut Grove's main entrance, Sheffield didn't think anything about the strong smell of smoke in the air. He figured it was from the car fire. Then, as he watched terrified men and women push their way outside, he realized there was something very wrong in the nightclub. At the same instant, the firefighters spotted a much larger problem than the auto fire.

With many of the Melody Lounge patrons carrying the fire on their clothing, the flames quickly spread to the foyer. Using the walls and decorations as a conduit, the fire literally burst into the main building. Perhaps fueled by alcohol fumes, the fire now seemed to be burning the air. Unbelievably, those in the rest of the club couldn't hear the screams of those fleeing the Melody Lounge. Dining, drinking and dancing was still going on as usual. As patrons continued to crowd around Buck Jones, no one knew panic was about to be spread by a wild fire.

Billy Payne and the chorus girls were climbing the stairs from their dressing rooms in the basement when he smelled the smoke. Rushing out into the dining room, the singer saw entire walls bursting into flames and shocked patrons suddenly racing from their seats and the ballroom floor. A burning piece of fabric from the wall came down and wrapped itself around two dancers who were unaware of the calamity that had taken over the club. Payne watched them burn as others fled. In seconds the path to the main exit was jammed with Boston's finest. Turning around, Payne begged those around him to follow him back downstairs. A dozen did, but most of the entertainers rushed out past the stage and into the mass of panicked people. They reasoned that their chances were far better there than they would be if locked in a basement with no exits. How wrong they were.

As a performer, Billy Payne had gained the reputation as a man who could think on his feet. As he flew back down the stairs and raced past the dressing room doors, he quickly and carefully assessed the situation. He knew that there was no escape upstairs or down. He knew that everything above him and around him was going to burn and burn quickly. The only chance he felt he and those who trusted him had was in the building's large walk-in freezer. Payne sensed the fire couldn't get into that room, but, as he opened the thick door and urged his friends inside, he wondered how long the air would hold out.

On the dance floor and dining room, most tried to get out the way they had come in, through the main door. They couldn't have known that it was jammed with terrified patrons and that the foyer was already a raging inferno. A few who took the time to assess the situation noted that many employees were headed in the opposite direction. Though they were hidden by drapes, there were exits on the back side of the room. Rather than fighting the mob, the observant few raced to the doors that opened on to Shawmut Street.

Within minutes of the fire spreading into the main part of the Grove, gasses built up in the club and people were fainting. As dark smoke moved down the hall and into the New Cocktail Lounge, another problem revealed itself. While the door here was not locked, it opened inward. Those fleeing for their lives from the flames pushed against the exit on the corner of Shawmut and Broadway only to have it remain solidly closed. As more and more frightened patrons stacked up at the door, opening it became impossible. As the fire flashed through the room, the exit to safety was jammed as tightly as was the revolving door in the foyer.

When those who had been hovering around Buck Jones saw the fire, the need for the cowboy's autograph suddenly seemed unimportant. Screaming in terror, they ran as fast as they could for any available exit. High on the terrace, Jones and his party had little chance. Fumes quickly hit them, bringing them to their knees. Smoke was so thick that most passed out within seconds. Though newspaper reports would later state that Jones carried several people to safety, this is highly doubtful. By the time the fire hit the main dining room, few people were getting out of the building and no one was getting in. It is more likely that the movie star passed out and fell to the floor before he realized just how grave the situation really was. This did not make the cowboy star a coward, only another helpless victim.

The firemen who spotted the blaze as they worked on the car fire probably put out the first alarm for the Coconut Grove. That call was issued at 10:17, and by then the fire had been burning for at least seven minutes. Even as the firemen arrived on the scene, even as they heard hundreds screaming inside, there was little they could do. With the exits jammed by the dead and dying and windows long ago covered with thick glass bricks, there was no way to conduct rescue operations. All the firemen could do was issue several more alarms, calling out every fireman in the city to try to battle the blaze before it spread to other buildings. The horror of their helplessness to respond to the pleas of the dying caused many firemen to fall to their knees and wretch.

Within fifteen minutes of the first burst of flame, it was over. Thousands of gallons of water might have helped douse some of the flames, but in reality, the fire had simply run out of fuel. The decorations were gone and the building's bricks would not burn.

Using axes to break the thick glass bricks, firemen finally got inside the club. What they saw was like nothing they had seen before. The smoke-covered men were greeted by a grisly scene that defied comprehension. Among the mass of twisted and trampled bodies was a man burned beyond recognition still leaning casually against the bar, a couple in the middle of the dance floor, embracing in midstep, a teenage girl in a phone booth, the receiver melted against her ear, and a bartender with a partially mixed drink still resting in his hand. John O'Neil and his bride sat by the orchestra, never making it to their honeymoon. More than one hundred bodies were jammed against the New Cocktail Lounge's door. If the door had only opened out, then none of these people would have died. More than a hundred people, some stacked eight high, were pushed against the revolving doors. Again, with standard opening exits, these people too would have been saved. Others were piled up against locked exits. It was a horrible sight and one that didn't have to happen. Many firemen cried as they gently touched incinerated piles that had just moments before been people celebrating life.

Those who hadn't run may have had the best chance to live. Payne and his band were uninjured in the freezer. Others, who had climbed to the roof, found a way to other rooftops and away from the flames. And many others who had either passed out in their chairs or waited where they stood or sat for what they thought was inevitable death avoided becoming human torches. The fire had somehow bypassed them as it sped through the building. No one inside the building escaped without injury, however. Thus, a call went out for every ambulance in the county and beyond. Squad cars, private vehicles and trucks were used as well. Still, it would take hours to get every one of the injured and dying out of the club.

The living were rushed to Boston City Hospital at the rate of six a minute. There was little doctors could do for most. Like Buck Jones, they were simply too far gone. The bodies of those who died in the emergency room or on the way to the hospital were stacked in the hallways in order to make room for the survivors. Those who had died at the club were placed in a building next door. A few hours later, relatives were given access to three temporary morgues as they combed Boston and the hospitals looking for loved ones. Many never found them since more than a hundred bodies were so badly charred that traditional methods of identification were impossible.

Joseph Boratyn would never see another Holy Cross game—he was one who never got home. Many of the soldiers, sailors and marines would never have to worry about facing Hitler's war machine; they had gone down fighting for their lives in a seemingly harmless night spot. One man who did live was Stanley Tomaszewski.

Once the fire had been put out, Tomaszewski sought out a fireman to admit that he started the blaze with his match. Taken to the fire inspectors, the youth repeated his tale and showed the men where it had all happened. The boy's story sounded reasonable until other witnesses reported that Stanley's match had not been near the ceiling or the fake palm trees. So, if Tomaszewski's match did not begin the disaster, what did ignite the ceiling in the Melody Lounge at the instant the boy changed the lightbulb?

Inspectors thought it could have been a patron's cigarette or lighter that began the blazing inferno. Almost everyone in the Melody Lounge was smoking, and with all the drunken horseplay that was taking place, it would have been easy for a decoration to have been ignited. Some witnesses even felt that a palm tree might have been set on fire as a prank, just like someone had stolen the lightbulb.

Another theory centered on the Coconut Grove's electrical systems. The wiring had been installed very cheaply by an unlicensed electrician. It was substandard, but did it create the spark that created the killer fire? No one could say for sure.

Other wild and radical stories involved arson, but there was no evidence of foul play. Some even thought the air was so thick with alcohol fumes that the booze should be held responsible. Yet after carefully studying the wreckage and questioning all the witnesses they could find, the inquiry could state only that the fire was started by "undetermined origins."

Amazingly, while hundreds were injured in the Coconut Grove fire and hundreds more died that night in the club, no fire laws were broken. Boston didn't require fireproof decorations or sprinklers, and exits did not have to be clearly marked. Management hadn't planned for a disaster but they also hadn't violated any city codes. But because 492 people—more than could legally be in the building—had lost their lives in the fire, the city of Boston demanded that someone pay. Especially since most of the victims would have survived if exits had been clearly marked, unlocked and made to swing out.

The club owner, Barnett Welansky, was found guilty of several non-fire-related charges and sentenced to twelve to fifteen years in prison. The man who built the New Cocktail Lounge was found guilty of violating building codes and sentenced to two years. No one else was found at fault and civil suits were unsuccessful.

Because of World War II, the fire at the Coconut Grove was soon lost in the news. Other nightclubs welcomed those who needed to party or unwind. Back in Hollywood, Buck Jones was replaced by Roy Rogers, Rex Allen and many other singing cowboys. New draftees filled the ranks of the servicemen who had lost their lives in the fire. By Christmas, except for the families who had lost loved ones and those battling injuries, life was back to normal. Yet for those who witnessed what happened the Saturday after Thanksgiving in 1942 at the hottest night spot in Boston, for those who somehow lived through it, life would never be the same. Many would avoid crowded buildings for the rest of their lives, others would not sit down in a restaurant until they had located all the exits, and most would never fully drive the smell of smoke from their memories. November 28, 1942, was not only a night to remember, it was a night that some could never forget.

# 7

# The Tri-State Tornado 1925

We name hurricanes. Relatively rare, they are the marathoners of storms, sometimes lasting weeks each hurricane season. We chart their courses, rate their power and even get to know their unique personalities. Andrew, Buelah, Camille and hundreds of others, all christened following letters of the alphabet, evoke images of massive destruction long after the damage they created has been cleaned up. Hurricanes are the powerful giants, the heavyweight champions of the weather world. They are the kings of the storms. Perhaps that's why we give them names, attribute human qualities to nature's fury.

Tornadoes are different. They are the small sneaky sprinters whose lives can usually be charted in seconds or minutes. We don't name them. Most do relatively little damage. Sometimes there are hundreds in a single day, but we usually come to know only a scant few of them. We most often reference these storms by the places they strike, such as the tornado that struck Waco, the storm that hit Xenia or the twisters that dropped onto Oklahoma City. Yet one funnel was so different and so destructive that it was given a name to go along with its infamous history, a nightmare brought to whirling life. In the world of storms the Tri-State Tornado stands apart as the champion of all tornadoes.

The Tri-State Tornado was fast, but it was not just a twisting sprinter. This humongous storm was a marathoner—a cyclone that combined speed with power and an almost human determination. It rode the ground like a freight train and delivered death and mayhem like a B-52 bomber. To those it struck, the storm seemed possessed. It had a will to live unlike any other tornado on record. Some even called it demonic, and while it is unlikely that it was driven by pure evil, there is little doubt that it had a thirst for blood.

On March 17, 1925, a little more than twenty-four hours before the twister formed, the United States Weather Service forecasters looked ahead to the next day. To the experts of the era it appeared as if the Missouri Ozarks and the plains of Illinois and Indiana were going to have rather normal spring weather. Using the best information at hand, the meteorologists predicted there would be a few showers, but mostly sunshine with unusually warm temperatures. Of course these men had very little but instinct to go on. In the twenties, weather forecasters did not possess tools such as radar, satellites, detailed wind profiling and computer modeling. They made prognostications based on observation techniques that had been used for centuries. Even if they somehow spotted a tornado, the only way to warn someone was to point it out on the horizon and say, "Run for cover." On this March day, no one foresaw any possible threat of deadly weather.

Though the experts didn't forecast it, March 18, 1925, was to be a day of volatile weather. A strong low-pressure front over the Ozarks of Arkansas and Missouri had moved northeast. Ahead of this system lay unseasonably warm, humid air. Temperatures, which should have been in the forties, were in the seventies by lunch. Those who were experiencing this unusually humid day knew things weren't normal. As they joked about the peculiarly pleasant weather, they also had to wonder if there was a catch.

Those enjoying the sun but looking to the skies knew that clouds were gathering in the west, and it appeared as if it might just shower a little later in the day. Still, with the winds light, severe weather wasn't a concern. The rain came quicker than expected, but even as it gently fell and the skies grew darker, few surmised that a big change was brewing. And no one guessed how dramatic that change would be.

What even fewer knew was how moist warm air trapped beneath a stable layer of cold dry air creates conditions ripe for a tornado, a situation called an inversion. When the warm air on the bottom of this "sandwich" hits a front, it can punch through the cool stable air and spiral upward. Supplied by winds at different levels of the atmosphere, the rotating funnel gains power and speed. This is how the tornado that would tear through three states in 1925 was born.

Just after lunch, a couple of miles as the crow flies from Ellington, Missouri, a solitary farmer dressed in bib overalls and an old hat noted the shift in the wind as he went about his daily chores. It might have been raining, but he still had eggs to gather and livestock to feed. Pulling his hat down and avoiding the shower as best he could, he went about his long-practiced work routine with little concern for the moisture.

When the day suddenly darkened and the wind stopped altogether, his eyes were probably instinctively drawn toward the skies. His heart must also have jumped when he heard the snapping of branches behind him. Yet even if he did react in time to glance back toward the now blue-black western skies, it is doubtful he spied a telltale funnel. With his view of the horizon blocked by tall oaks and rolling hills, all he probably saw was a wall of falling water. By the time the rapidly moving sheet of rain pelted his face a few seconds later and a sudden burst of wind blew down his fences, it was too late. As if it had lassoed the man with an invisible rope, the storm picked him up, shook him like a rag doll and violently tossed his limp body to the ground. He never knew what hit him. He never had time to look back on his hard life in the fields or think of his family and home just over the rise. With unmerciful power, the Tri-State, though not yet roaring like a train, quickly crushed the lonely man and left him buried in the Missouri mud. It was just a small foreshadowing of the destruction that would mark the rest of the afternoon.

If that farmer had been the only victim, then the tornado would have quickly been forgotten, deemed nothing more than a freak wind. This storm, however, seemed to possess an unquenchable blood lust. One victim didn't satisfy it. It wasn't appeased by simply tearing up a farmer's plowed ground. It seemed bent on hitting any and every population center that came within its dark grasp.

For the next three and a half hours, more people would die and more schools would be destroyed, more students, businessmen, housewives and farmers would be killed or injured than from any other single tornado in U.S. history. This massive storm would set records for speed and path length that will probably never be broken. This was a giant born to be a killer and its determined rage made it unforgettable.

As if single-minded, the Tri-State tornado maintained an exact heading, N 69 degrees E, for almost 200 of its 219 miles. As if targeting the small towns that had been built along a railroad line, it struck with fury, spinning on its direct course at more than a mile a minute. No car of the time could have outrun it, nor could many have driven around it. For more than one hundred miles of its destructive path, the storm's width held at least three-quarters of a mile. And with only a few rural observers noting its birth, there was no way to warn people in the storm's way that death was rushing toward them in the form of the meteorological equivalent of a mass executioner.

"What's that, Daddy?" a five-year-old girl asked her father as they walked through the family's fields. It was less than two minutes since the Tri-State had claimed its first victim.

As the man studied the dark mass of clouds he answered, "Looks to be a storm brewin'. We best get back home."

"I know it's a storm, Daddy," she replied. Then pointing her finger toward the clouds, she continued, "There is a big tree up there in those clouds. How did it get there?"

The man didn't take the time to answer. Picking his little girl up in his arms, he rushed back to his cellar. He knew too well what the folks just a few miles away were about to experience.

Annapolis, Missouri, was a hill town. It was as far removed from its namesake, the home of the Naval Academy, as it was from the ocean. Hidden in the midst of tall stands of trees and rock-covered hills, the community of 350 was home to humble people all but untouched by the roaring twenties and the passage of time. Children and adults alike still marveled at automobiles and craned their heads upward anytime an airplane skated across the heavens. Families lived in frame houses they usually had built themselves. Few had phones and many didn't have electricity. If Jesse James or Mark Twain had come back to life and found themselves on Annapolis's dirt streets, they wouldn't have noted much in this hamlet that was very different from the time they called Missouri home decades before. It was quaint, neat and very quiet.

At about one o'clock, as men and women went back to work after their lunch breaks, the light skies suddenly darkened. The operator of the general store noted a sudden ear-shattering blast of thunder. Strolling out the door he had opened earlier to let the unseasonably warm breeze filter through the building, he took a look at the horizon. Like others who looked out their windows, he noted a gathering darkness to the southwest, but passed it off as a low-lying fog bank. Nothing to worry about, he thought, it wasn't even raining that hard.

The blackness on the horizon was hardly harmless mist, it was a swirling mass of power almost a half mile across. Within a minute of being spotted but remaining unrecognized the tornado hit Annapolis with hundred-mile-an-hour winds and raindrops the size of marbles. In seconds, the community was ripped asunder, and as frightened survivors clawed their way out of the wreckage, they all wondered what had just happened. They were surrounded by bricks, uprooted trees and piles of debris that used to be buildings, but no one ever saw a funnel cloud.

As rain poured down off men's hats and women's bonnets, as children cried and shocked animals walked warily down now-muddy streets, the reality of what had just transpired began to sink in. Ninety percent of the town was in rubble. Block after block of Annapolis had been leveled, just a few foundation stones left to mark where the town had stood. Debris was scattered everywhere. Papers, books and clothing were caught in trees and homes had vanished. It was as if they had never been there at all. And scores of people were hurt, bloodied, bruised and confused.

The mining town of Leadanna was two miles to the northeast of Annapolis. It was so small that there weren't even population signs posted on the road. Everyone knew everyone and most worked in the area's iron mines. Less than two minutes after it had plowed through a stunned and unprepared Annapolis, the Tri-State hit its neighbor. This time a few farmers saw not one but two funnels. Because of the trees and hills, however, they couldn't gauge either's size or speed. They knew the storm was powerful because they could see pieces of tin and boards swirling along with the dirt that was rushing up to meet the sky. They could also tell that Leadanna was directly in its path. And there was nothing they could do to warn the residents.

Passing through the hamlet even more quickly than it had Annapolis, the now roaring storm damaged or destroyed every building in town. Two died, hundreds were injured. The wreckage from the two communities somehow meshed together in haphazard piles, deposited in trees, along roads and draped like confetti over the top of the hills. The funnel hit a school, where students were thrown violently out of the building and onto the playground. Men and women were ejected from their homes. Animals were caught up in the winds and carried for miles. For a few seconds, hail as large as baseballs pelted the ground and everything and everyone on it. And then it was over. An eerie silence settled in. The Tri-State had moved on as quickly as it had appeared.

Seconds after the roar became just the pattering of rain on the ground, a child, bleeding from his head, picked up a soggy picture of a family he didn't know. Wiping away the mud, he stared at the image. Then he began crying out for his own mother. Oblivious to the destruction that was all around him, he wanted to show her his prize. But the woman couldn't be found. Even if the toddler had found her, she might not have recognized him or the photo. Most of the community's residents were either in shock or badly injured.

Slowly, Leadanna residents followed the tiny boy's lead and pulled themselves out of the wreckage. Most still hadn't guessed what had hit them. It had all happened so quickly. They hadn't seen the storm coming. Except for the sky suddenly growing dark, there had been no warning. Yet as they looked at the smashed buildings that had once offered them security and the piles of lumber that had been their homes, they began to guess that this destruction must have been caused by what they called a cyclone. A few blamed the devil himself.

If the storm had ended there, with a toll of three lives and a few hundred injured, it would have only been remembered for wiping out Annapolis and Leadanna. Leveling the two Ozark hamlets would have made for a powerful legacy, but not one that would have placed it in the record books. The twister was far from satisfied; what it consumed in Reynolds and Iron Counties were nothing more than appetizers. The main course was still ahead.

For the next half hour, as the Tri-State moved across Madison County at more than sixty miles an hour, it re-formed into one funnel and stayed glued to the ground. Miraculously, while it plowed up fields, mowed down trees and killed and maimed countless animals, it didn't come in contact with any people. Once out of the Mark Twain National Forest and into the Mississippi River bottomland, the twister grew nasty. In Bollinger and Perry Counties, the death toll quickly mounted.

Though modern educational reform was sweeping the East Coast, most rural schools in the hills of Missouri were still taught by a single teacher with as many as eight grades sharing a lone room. While it may not have been an ideal way to prepare children for a rapidly changing world, there was a sense of security that was a part of this system. The older kids helped the younger ones, the teacher was as much a friend as an instructor, and the school was the glue that held together rural families who had no phones or cars or radios. Yet on this day, the security that was found in the small schoolhouses would be forever altered.

The kids studying their arithmetic heard the wind pick up and watched as the thick clouds plunged one single-room Madison County school into a surreal darkness. As they glanced over to the windows, they noted that the rain began to pound on the thin glass panes. The thought of having to walk home in the downpour sent a collective shiver through the room. With huge drops now peppering the tin roof and completely drowning out the teacher's voice, a couple of the older boys ventured over to the windows. In the eerie bluish-green haze, they noted a dark mass. Before they could react or warn the others, that mass hit the school with its full force. In seconds, the old building was turned into toothpicks. Amid the rumble of boards, broken desks and school books more than a dozen kids moaned, their shocked expressions reflecting the suddenness of what had hit them. Most had no idea where they were or what had happened. A few were trapped under sections of roof and pieces of shattered floor. Most had been tossed like wadded-up scraps of paper across the schoolyard. With one dead and twelve victims crying in pain, the storm sought out two more Bollinger County schools in the next twenty minutes. The destructive nightmare was repeated again, but miraculously only one child died and three dozen were injured.

When the storm raged into Perry County, it again broke into two massive funnels. As if fueled by the terror they created, these twin twisters began picking up trees, houses, and wagons and carrying them in their swirling masses for miles. Along the way, a few more victims found themselves caught up in the wreckage. Scores were injured and maimed, a few more were picked up and then violently dashed to the ground. With an anger usually reserved for creatures of flesh and blood and a purpose almost always lumped together with a demon, the storm pushed on, now howling like a mad wolf as it set its sights on the mighty Mississippi River. Splashing into the river, the Tri-State merged again into a single funnel and waved good-bye to the Show-Me State and the thirteen unlucky men, women and children it had killed there. As would soon be obvious, despite sixty miles of leveled ground and bleeding victims, Missouri got off lucky.

No one saw the storm take on the river, but as it raced across the mile-wide section of water in less than a minute, driftwood, fish and snakes were caught up in its whirling clouds. The storm was now so large that for an instant it must have spanned the Mississippi River like a bridge, its sides touching both the Missouri and Illinois banks. But this was not a bridge that anyone could have safely crossed; if anything, the Tri-State was a bridge connecting life to death, reality to nightmares.

As it began its trip across Southern Illinois, the tornado grew more powerful. Now sporting winds of more than two hundred miles an hour, the twister was a raging mass of unyielding power. Though the area in its immediate path was largely unpopulated, the storm seemed to thirst for human contact. Ignoring the wide-open spaces just to its left and right, it set its sights on tiny Gorman, Illinois. The small town, completely unprepared, had no idea that a monster was just moments away.

Gorman had five hundred residents who depended upon the river and farming to sustain what was a hard living. The town's folk had fought a driving rain all morning, but by afternoon the precipitation had slowed to a drizzle. Still, even as the rain slowed and the sun appeared, the damp gloom hung on as if foreshadowing what was just ahead. At 2:30 it seemed as if something had suddenly blotted out all signs of light. In an instant it had become as dark as midnight and as still as a tomb. Rushing to windows, Gorman's citizens looked toward the southwest. They couldn't see a definable funnel cloud, but a black-green mass of clouds hovered on the ground and was growing closer with each passing moment. Transfixed by a sight that seemed neither harmless nor frightening, many simply stared at the strange phenomenon. Then, as it hit the outskirts of town and pulled trash cans, fence posts, wagons and boards into the air, people realized that this was more than an unusual bank of uniquely beautiful clouds.

At the Gorman Cafe, a few folks were finishing a late lunch when the storm hit. Rain began pelting the roof and the clouds made the day pitch black. The patrons found their eyes drawn to the city's main street. There they saw wagons rolling by without horses or drivers, milk cans and cook stoves flying just inches off the ground. Then, as the lunch patrons began to realize the seriousness of the situation, a thousand-pound cow, bellowing in fear, flew down the street and smashed into the building's front wall. Panic immediately set in.

Running from the storm's path, a mother and baby rushed out into the street. The tornado ignored the woman but reached out for the child. Frantically the woman pulled against the twister, but its two-hundred-mile-an-hour winds would not give up. Wrestling the baby from her arms, the storm lifted the child airborne, now surrounded by homes, animals, bricks and thousands of other common household and farm implements. The scene looked like a huge mobile. But it was also a deadly one, violently snuffing the life out of a baby who had not yet learned to express his fear.

At the city's main school, the children rushed from their seats to the windows as soon as they heard the howling wind. As they looked out into a day that had become dark as night, teachers frantically tried to get them back into their seats. They never had the chance. The old school collapsed under the wind's fury and the floor, desks and children were buried under bricks, roofing materials and broken glass. The pounding rain and swirling wind muted their cries. Seven of the children died before their families could begin digging through the rubble.

The storm raced across the hamlet, taking out block after block, picking up homes as if they were dollhouses, often carrying them and their residents hundreds of yards before slamming them back to earth. In the brief span of two minutes, almost all of Gorman was wiped off the face of the earth. It looked more like a trash heap than a town. Sections that had once been the jewels of the community were now nothing more than open prairie. There were no trees, no homes, no fences, just the outlines of foundations and a few clocks stopped at 2:35.

More than half of Gorman's residents were injured. Thirty-four were killed. Yet in the moments after the storm's rage, the sun came out and a strange silence fell over the town. Shocked residents, some stripped naked by the winds, staggered like punch-drunk boxers down streets, not seeing the destruction or feeling their own pain. Numbed by a power that had left scores of them mute, they were waiting to wake up from a nightmare that had somehow crossed from their dreams into reality. It would be more than six hours before anyone from outside the community heard about Gorman's plight and could offer help. Even at that, many were still consumed by a paralyzing shock.

Some in Gorman began to grasp what had happened and forced themselves to search for victims, but the communities just ahead of the storm were still going on about their normal lives. With no modern means of communication or storm detection, they had no idea of what awaited them.

Murphysboro is located ten miles northeast of Gorman. Bustling with life, it was the mecca for goods and services in this part of the state. Rare for the area, Murphysboro had industry. Isco Baste, Mobile and Ohio Railroad and Brown Shoes all had large operations in the town. Because of the hundreds of jobs these facilities offered, this was a community looking to the future with pride, banking on tomorrow being better than today. But as it turned out, the residents should all have been looking toward the clouds and not to the future, because the approaching storm would forever derail Murphysboro's hopes and dreams.

The first reported Murphysboro sighting happened just before the storm hit town. A few farm families saw what looked like large sheets of paper floating in the sky. This was just the leading edge of the twister and was really a false front that hid the deadly power that was seconds behind. As had been the case in Missouri, the ground-level boiling mass of clouds and the massive amount of debris caught in its grasp hid the funnel. It was as if the Tri-State was disguising itself in order to surprise more prey. To all who observed the twister that day it looked like nothing more than a fast-moving thunderstorm. Because of this illusion, few took cover. This would prove to be a death sentence for hundreds.

In Murphysboro, the Tri-State terror not only struck, it turned the entire fabric of the community inside out. Rushing right down the main highway with winds that exceeded 175 miles an hour, it roared through the town like an Illinois Central express train. Effortlessly tossing bricks, trees and cars in every direction, the twister destroyed everything in its path. On this dark day more people would die in Murphysboro than in any town ever struck by a single tornado. The 234 deaths unfortunately included at least 25 children in three different schools.

While Murphysboro was a growing town in step with the industrial age, its schools were built in a style that harkened back hundreds of years. Constructed of stone and brick, they had no reinforcement. The tornado blew through the buildings as if they were made of paper, tossing the crumbling walls in every direction. Children had bricks and school books pelt them again and again. Many were buried under tons of debris from falling walls. Few escaped unhurt.

Three factories, the pride of Murphysboro, quickly contributed to the second fatal blow to the already bleeding community. The tornado had destroyed the complexes and spread the fire from their furnaces across the town. With the power plant among the buildings damaged by the twister and with the water-supply system destroyed as well, there was no way to fight the fires.

A few seemingly lucky souls had seen the storm in enough time to run to their basements. They survived the twister, but they were trapped under the pieces of what used to be their homes. As they counted their blessings and patiently waited to be dug out, the fires spread above them. Many of these people's refuges of safety would become ovens of death. With no water, all that friends and family could do was listen helplessly to their screams as flames that were visible for sixty miles enveloped home after home in the town.

Murphysboro's losses totaled more than ten million dollars, a staggering figure for 1925. Half of the community's population was either killed or injured, forty percent of the homes were destroyed, thirty percent were damaged by strong winds and fires and a hundred city blocks were simply erased. And the trio of industries that had been such a source of community pride would never reopen.

After Murphysboro, the Tri-State tossed boards, bricks, cars, wagons and furniture across the landscape. One house was picked up, taken a hundred feet in the air, twisted around, then exploded into a thousand pieces not much bigger than rulers. As if on a mission, the twister continued on its track, headed toward the tiny village of Desoto.

If Murphysboro was on the cutting edge of tomorrow, a town with a bright future, Desoto was really just a group of houses and a few businesses trying to hang on to the moment. The main commercial business district consisted of just two city blocks. The community had gotten electricity only a few years before and some old-timers still used oil lamps and wood stoves. No one knew what had happened southwest of them just a few minutes before. Few were looking at the skies and thinking that they were going to be hit by anything other than another shower.

Those who heard the Tri-State coming noted the loud roar but didn't see cause for alarm. Desoto was accustomed to freight trains. It was a mining town, and coal train noise was easily dismissed. So no one looked out to see what all the roar was about. Even if someone had, what could they have done?

The Tri-State and its mile-wide mass of destruction hit Desoto like a bomb. It literally blew up the town. It sucked women and children out of their homes and into the air and beat them to death. Some kids had been dismissed early from school so they could get home before it rained. They were walking home when the storm hit. One elementary school girl was picked up and wrapped around a fence. Others were so badly disfigured that they could only be identified by the color of their hair or the clothes they were wearing. The few who did survive their short trips home were bruised and cut by flying masses of debris that defied description. They were mentally and physically scarred forever.

Those who were caught in the storm walking home from school might have been lucky. Of the 125 students in the Desoto school, 33 died as the storm directly hit the old school building.

The storm raged on, eventually killing sixty-nine people in the small hamlet while most of the men of Desoto worked in ignorance in a coal mine. Only when the shaft lights flickered and then went out did they sense that something was wrong. When the power didn't come back on, the miners climbed one by one up the escape shaft and were greeted by a scene they hardly recognized. Their town was gone, destroyed. Bodies and pieces of bodies, along with countless dead animals, bricks, stacks of broken lumber and thousands of shattered household items littered the streets of their town.

With no one to organize a rescue effort, husbands raced to their homes and the school. Many found their wives holding dead babies, trying to urge the crushed little ones back to life. Others dug with their bare hands, pulling children, some dead, some alive, from the piles of bricks and boards. Yet as the rain continued to fall and the temperature dropped, the hopelessness of the situation began to hit those left alive. Five hundred healthy men, women and children had called Desoto home. Within a few seconds, fifteen percent of them had been killed and two-thirds of those left alive were injured. It would be hours before outside medical aid arrived to try to help the community come to grips with what the Tri-State had done in less than a minute. It would take mass graves to bury all the victims. Desoto would never be the same.

Now moving at sixty-three miles an hour, the storm continued on its direct course across southern Illinois. Spitting out pieces of Desoto along its way, it bore down on the town of Hurst. In seconds it ripped the tiny town to the ground, killing seven and injuring twenty-seven more. The town's few residents never knew what hit them. The storm didn't stop to explain; it had much larger prey in its sights.

West Frankfort was located ten miles up the road. It was a growing community whose people depended upon the rich coal deposits in the hills around the town. Eight hundred miners who called the city home worked dawn to dusk six days a week for three different mining companies. As the Tri-State roared along the Illinois earth, they were all safe five hundred feet below the ground. Yet several thousand others, most related to those who worked beneath the brown soil, were not so lucky. They were directly in harm's way.

The twister had now slowed to fifty-six miles an hour, but it had also spread out. It was a mile wide as it approached West Frankfort. Again as if hiding its intent, the storm produced no lightning and lightened in color as it bore down on the community. Several people were taking advantage of the unusually warm weather by working in their yards when they noted its approach. Almost all of them stopped their labors long enough to stare at the mustard-colored mass of clouds that were clinging to the ground and rolling toward them from the southwest. The clouds were odd but beautiful, unlike any that anyone had ever seen. Yet they did not appear menacing. They didn't create concern or terror, only curiosity.

The storm bypassed the town's business district, heading instead toward the miners' homes. Mini-explosions preceded the hidden funnel along the streets as family after family's hopes, dreams, security, and in many cases lives, were lost in the swirling mass of clouds. The power instantly died, as did more than one hundred people. By the time the storm raced through the last home in its path, a fifth of the community had been wiped off the landscape.

It took the miners more than half an hour to struggle up from their pits, and what they found spread out before them was unlike anything they had ever seen. Even those who had served in World War I couldn't fathom this kind of mass destruction. As the shocked workers wandered the streets, the rain had stopped, the wind had shifted and there were no signs of the killer who had struck with such viciousness a few moments before.

The few doctors in town had to treat men, women and children where they found them. Concern about gangrene caused the physicians to amputate broken limbs without the aid of any pain-killing drugs. Screams were heard from almost every pile of wreckage and grown men could be seen crying on their knees in the middle of lots that used to be homes.

A young girl was holding a baby and walking from person to person asking them if they knew who the child belonged to. She discovered it on the street and she wanted to take it home. She didn't know it was dead.

For a brief moment the sun came out, bathing West Frankfort in a radiant light. A few people looked up from their rescue efforts long enough to note the appearance of a beautiful rainbow. Yet no one stopped their labor to search for the pot of gold. No one tried to find any silver linings on this day either.

The Tri-State killed 148 people in West Frankfort. Another 400 were seriously injured. It could have been much worse, though. If the storm had hit in the center of town, more people surely would have perished. If it had struck after dark, the numbers would have grown as well. But it was hard for the residents of this town to take solace in those facts when more than twenty percent of its people were now dead.

For hours, rescuers, mostly miners, were confronted with gravely injured people who didn't know they were hurt. They were somehow unaware that a broom stick was embedded in their side or a piece of lumber had been driven into their heads. Then, in a bizarre twist that was lost on no one, snow began to fall. Surely, many thought as the death toll mounted, this was a day that was too strange to have really happened; it had to have been just a bad dream. One that continued in town after town.

Caldwell was not really a town, but a junction. The folks who lived in the area claimed Eighteen as their home. Eighteen didn't have a post office, it was just a tiny school and a group of small modest homes. Classes were just ending when the children heard the storm's roar. As in other communities that had been touched by the Tri-State, most thought the noise was just another coal train. Yet as the large drops of rain began pelting the building, the teacher decided to keep the kids in class until the rain abated. It wasn't long until the children were sent on their way, completely unaware of what had struck just beyond their playground. What they found were fields where their homes had once been. One of them spotted a woman and her baby in a swamp along the road. Another came across three generations of the Karnes family, all dead. They had been blown out of their home, tossed on the hard dirt after flying through windows and walls.

In Eighteen, the few homes that survived were quickly turned into hospitals or morgues. The roads were all blocked by debris, and there was no way to contact anyone in the outside world. The only ones who could try to help the injured were the children from school and the few adults who somehow had escaped the storm's wrath. Their good intentions proved too little and were too late to help the more severely injured.

At 3:15, just over two hours since the Tri-State had claimed its first victim, it roared into Parrish. There were few men to greet it. As in Desoto and West Frankfort, they were deep in the coal mines. It was the women and children who were left to face the storm.

Parrish had three hundred people who lived in about fifty modest houses. There were a couple of churches, a post office, a school and a small business district that held the community together. Seemingly intent on claiming everything, in less than a minute the Tri-State destroyed all but one home, the school and a church. Its winds now in excess of 250 miles an hour, the tornado obliterated whatever stood in its path. It didn't just damage structures, it carried them away, not leaving even a trace on the ground. Nothing was spared. The postmaster was blown from his office, then wrapped in electrical wires. Sixty children huddled in their school, scared but unhurt as their homes and families literally flew past them. Then, as quickly as it roared into town, it was gone.

Those who lived through the minute of unholy terror and could manage to get to their feet had only a moment to consider what had transpired. Then baseball-sized hail began to beat down on them from the heavens. With few structures left for shelter, most grabbed tubs or parts of roofs to fend off the new enemy. The relentless pounding continued for almost five minutes. It was only then they could begin to uncover the twenty-two people who had died and the more than one hundred who were injured. After the mass funerals, those who lived in Parrish moved away rather than rebuild.

The tornado now found itself in the farming belt of southeastern Illinois. It was moving at almost seventy miles an hour. Though there were no towns in its path, people still felt its wrath. Terror reigned as the normally weather-wise farmers were caught unaware of what was bearing down on them. With such a great forward speed, using its boiling mass of clouds to hide its mile-wide funnel, the tornado gave these people no time to react. Sixty-five in Hamilton and White Counties died. There were single deaths in three different rural schools. Beyond the toll in human life, the twister tore up fields, barns, homes, fences and spread a collection of two hundred miles of trash along its destructive path. As it approached the famed Wabash River, the dividing line between Illinois and Indiana, the Tri-State had already claimed more than six hundred lives and was bent on seeking out more victims.

After drinking in the water from the river, the storm moved toward the tiny town of Griffin. Again, with no funnel visible, no one could see the power directly in front of them. To most, the now greenish-blue clouds looked too pretty to be deadly. Still moving at more than seventy miles an hour, the storm's width had shrunk to three-quarters of a mile. A lot of good that did the people of Griffin. The Tri-State was still large enough to engulf the entire village.

The tornado struck the community with the violence of a full F-6, the most powerful rating on the storm scale for tornadoes. In the process, it leveled almost everything in its path. Children in school buses were sent flying through the air, old cars and farm implements were picked up and used to smash fleeing townsfolk. Death was severely and randomly dealt out in three-hundred-mile-an-hour wind bursts. Those who survived were not only injured but completely caked with mud. All 150 buildings in the town were decimated. Nothing was left standing. Half of the ninety who died in the storm's Indiana rampage died here.

If the Tri-State had held to its determined path as it had since its birth, there would have been no cities left for the storm to torture and few, if any, lives would have been taken in the next twenty minutes. Yet the storm had a mind of its own. It made a nine-degree turn to the north, thus putting it right on the path to Princeton.

As it raced along the southern Indiana landscape, the tornado appeared bored. Not content to simply tear up farms and terrorize rural families with its powerful winds and its three-quarter-mile-wide funnel, it decided to show off by splitting into three different vortexes. As the three funnels, now clearly visible, ripped through houses and barns, the sheer passion of its destructive nature revealed itself. The Tri-State picked up trees, buildings and animals and literally passed them from funnel to funnel. Those who watched in frightened awe felt the storm really was the devil incarnate.

After destroying eighty-five farms, the Tri-State's funnels again merged as it prepared to hit Princeton. Still moving at more than seventy miles an hour, the twister first struck the home of Richard Walter, killing three generations of that family. Next it killed two farming brothers, William and Walter King, and their wives. Then it moved into the town, murdering several dozen citizens, injuring hundreds and destroying more than half of the community's homes. For several more miles the massive storm continued to roar and tear up the landscape. Then, unobserved, like a thief in the night, it pulled back into the sky, never to be seen again.

In its wake, the Tri-State Tornado completely destroyed four towns, severely damaged six more, destroyed fifteen thousand homes, and killed 695 people. It stayed on the ground for 219 miles at an average speed of sixty-two miles an hour. In an ironic end to its reign of terror, the Tri-State, one of the few tornadoes ever to be known by name, left an unknown baby on the streets of the last town it struck. No one in Princeton knew the dead child, no one knew where it came from, only that it had been dropped from the sky by a storm so deadly and powerful that it seemed to have a god-like control over who lived and who died.

A marathon killer disguised as a tornado, the Tri-State struck fear across America's heartland. Even though few who felt its wrath firsthand are still alive to share their stories, dark clouds are still feared and images of death and destruction still remain fresh along the twister's trail. Four generations later, the Tri-State still lingers in the minds and fears of those who live in its wake, and few who live along this path ever feel completely safe when there are dark clouds on the horizon.

# 8

## The Chicago School Fire 1958

On December 2, 1958, a wire photo of a fireman carrying a boy shocked the world. This single picture of Chicago firefighter Richard Scheidt grimly taking a dead child away from a burned-out shell of an elementary school shook the very fiber of millions of parents and children in not only the Windy City but in London, Moscow and Tokyo. It was just one scene, one small fragment of a tragedy too cruel to consider, but this picture literally caused the world to pause, to pray and to ask, "How could this happen?"

In truth, it had to happen. It was a wonder it had not happened before. It was long overdue. In 1958, in New York City alone there were more than one hundred school fires. Miraculously, none had become anything more than a single-paragraph story buried in the backs of newspapers. In the United States, thousands of schools had experienced minor fires over the previous years. In spite of ancient buildings, overcrowded classrooms and century-old fire prevention systems, Americans had somehow dodged the one big fire event that would educate the nation as to the condition of thousands of its schools. For the moment, people could worry about the bomb and the Communist Bloc, not the flicker of flames in buildings that housed their loved ones. But time was running out.

Located in the heart of Chicago, Our Lady of the Angels Elementary School was like many of the schools parents sent their children to each day. With its red bricks and multistory framework, it gave the impression of great strength and security. Its hallways were wide, its floors and desks worn by generations of children, its big doors strong and welcoming. Impressions were shaded by more fiction than fact, however. The school was really a tinderbox, a fire waiting to happen. Every fireman who toured the facility knew that this school and hundreds of others in the area were really unsafe. Still each assured the other that it couldn't—wouldn't—happen here. God wouldn't allow it.

God might just have been a part of it, though. Because of the nation's rigid policy of separation of church and state, city inspectors ignored many deadly situations at the Catholic schools. In Chicago, where more than a quarter million children attended Catholic schools, giving the church some leeway seemed a natural. Much of the city was Catholic, and the church's schools were well-known and respected for their discipline and structure. They were as much a part of the fabric of the area as were the Cubs, the White Sox and the Bears. Most of the nuns who taught the classes might have been strict, but they were also as popular as Ernie Banks or Nellie Fox. The Catholic schools were an integral facet of the city's history and life.

In Chicago, each Catholic school was attached to a local parish church. Families would send their children to a school and usually attend services at the affiliated church as well. So each of these centers of worship and education were embraced as an important piece of a neighborhood's identity. Only a few miles from the city's famed Loop, there was nothing about Our Lady of the Angels, OLA to the locals, that really set it apart from scores of other parochial centers of learning and worship. To those who called it home, OLA was the center of their lives.

The school's main building was built at the turn of the twentieth century, with other additions completed on an as-needed basis. Our Lady of the Angels' framework and flooring were all wood, dried through the years. Steel had not been used in any part of the construction. But again, this was not unusual for public buildings constructed in the first half of the 1900s. Most of the stores and homes that surrounded the facility were the same type of construction. Yet as the city grew and more and more students were welcomed into the school, times began to change. Even though no one realized it, OLA silently became a much more dangerous place.

As crime levels rose in the neighborhood, the church diocese had approved construction of seven-foot-high iron fences around the playgrounds. As the baby boom crested, more people crowded into the area, thus creating a larger base of children who needed education; storerooms and offices became classrooms. By the early fifties, classrooms that had once been home to twenty students and a teacher now often housed more than sixty. Still, only one nun was assigned to each room. With desks pushed against each other and aisles almost nonexistent, OLA, like many other city schools, had become a claustrophobic place where students sat practically shoulder to shoulder in every room. It had happened so gradually that few noticed what should have been so obvious. The school and its staff were overextended beyond reason.

In 1958, the school's enrollment grew to more than sixteen hundred students. That increase forced a number of changes to the building itself. Supplies, trash and extra books were now housed under stairwells and in the boiler room. Though the staff attempted to keep things neat and orderly, most items and stacks were just pushed into any empty area. These stacks and piles often hid items both teachers and students needed. It often became a treasure hunt to find something. Anyone could see that the mess, caused simply by overcrowding, had created not only a storage problem but a fire hazard. City inspectors, however, continued to give the school passing marks upon every one of their visits. And though the school talked of addressing the space problems, they hadn't formulated any firm plans.

The city gave OLA a positive rating in part because government did not want to butt into the way that religious groups ran their own institutions. If OLA was a dangerous firetrap or if the students were not being taught well, then Chicago officials figured that the Catholic church leaders should and would handle it. Besides, the schools and churches were under the direction of the powerful Archdiocese of Chicago and not a part of the Chicago Board of Education. Even if the school had more than twice the number of students it should have had, it was easy for the city to ignore it. After all, the public schools in the city were also overcrowded and certainly didn't have room for those who attended OLA. So, while it wasn't an ideal situation, it was one that had always worked.

The area around Our Lady of the Angels was made up of large pockets of recently relocated European Catholic immigrants. Like many of Chicago's neighborhoods, the community was set apart from other areas of town by its citizens' unique culture. People here often spoke in their native tongues. Many of their names were considered long and hard to pronounce. They shopped for special foods and clothing and had holiday celebrations that most longtime Midwesterners didn't understand. Though new to America, these people quickly became proud of the lives they were carving out for themselves. And the focal point of this life was the church and school. That block, on the corner of Iowa Street and North Avers Avenue, was the center of their world.

On December 1, 1958, mothers who packed lunches and pressed clothes for their children were hardly worried about what would happen to them at school. It would be a cold walk, with a wind chill close to zero, but it would be worth it. With the coming holidays, school would offer their kids a chance to learn and to share in the fun of preparing for the season. Special programs were ahead, as were art classes spent making gifts for relatives. In so many ways it was the very best time to be in school.

All around OLA, the street's shops were putting up Christmas decorations and children spent their spare time looking in windows and making long wish lists. As in most close-knit neighborhoods, many of the kids would meet on the sidewalks and stroll to school together, sharing tales of their weekends and dreams for the holidays. Their noses, fingers and toes might have been cold by the time they arrived at OLA, but they all knew that John Raymond, the school's janitor, would have their classrooms warm and toasty long before the first bell rang. On this sunny morning, this picture seemed to be a snapshot of what was right and wonderful about the American way of life.

As the children left home, the nuns who essentially ran the school met in the chapel. Wearing the huge square hoods that identified them as a part of the Blessed Virgin Mary order, the women had a unique ritual they followed each day. As they gathered for prayer, they never failed to ask for protection against fire. The order's convent had burned to the ground more than seventy years before and it was something that the nuns were not allowed to forget. Since they had all heard the horrible stories and knew that they did not want to have to face what their founders had faced in Iowa, these prayers for safety were always heartfelt and sincere.

Students and teachers alike hit the doors of OLA just before eight. About half the school's students walked through the doors that led to the north wing. Once inside, they fanned out either into the ten-foot-wide middle hallway on the first floor or climbed wide wooden stairs to the second floor. These old stairs, worn by hundreds of thousands of small feet, dipped in the middle, and the noise that the children's leather shoes made as they hustled up and down the old dried wood echoed off the ceilings and into the classrooms.

The stairs emptied out into the second-floor hallway. Unlike on the first floor, students didn't have to walk through a fire door to get to the hallway or their six classrooms. Everything was open up to the twelve-foot ceilings. It made the school seem less crowded and gave the nuns a much clearer view of student activities, but it was also an important ingredient of the formula for tragedy. At the far end of the hallway, all but forgotten and dismissed as nonessential, was a second set of stairs. The only time the door to these stairs was ever unlocked was during a fire drill, so few students or nuns really ever thought of them.

As the students entered their rooms, few noted the fire warning switches on the hallway walls. Six feet above the floor, almost out of reach of some of the shorter nuns, these red boxes assured parents and teachers alike that if a fire did break out, a warning could be sounded in time to get the children to safety. Yet the boxes only notified those in the school. They were not connected to the local fire stations. In addition, if the fire was in the hallway, the nuns could not have reached the boxes, nor could the children have exited the rooms. With only one fire escape and the second stairwell locked, most had no safe path if the unspeakable should happen.

On this Monday, a flu outbreak had kept home about ten percent of OLA's students, though a few of the sick ones did straggle in by lunch. Still, the rooms were filled to overflowing. The morning and early afternoon went as usual for the children who were dressed in blue and white uniforms, paying strict attention to the nuns. By 2:00, with the final bell only an hour away, the teachers were having a difficult time keeping the kids' attention.

With only one janitor assigned to the maze of classrooms, nuns' quarters, priests' residence and the church, students were often called upon to help keep OLA looking neat. Each day several groups of boys from each classroom were assigned the task of emptying trash cans into the large waste containers in the basement. While he appreciated the help, on several occasions janitor John Raymond had reported problems. He often found cigarette butts in the boiler room and under the stairs. It seemed that a few of the boys had used the moments away from supervision as a chance to smoke. Though warned that anyone who was caught smoking would pay a stiff price, the butts still appeared from time to time near the waste cans.

The last boys who gathered and dumped trash reported back to their classrooms around 2:30. None reported anything unusual. Though apparently no one noticed it, one of the trash cans in the closed-up boiler room beneath the stairs was smoldering. With the doors closed and no oxygen to feed it, the flames did not immediately escape their point of origin. The fire, combined with the heat from the furnace, began to sharply raise the temperature in the confined area. Soon it was sauna-like. By 2:35, the small fire became hot enough that an old single-paned window in the room cracked and fell out of its frame. This allowed a Lake Michigan–driven north wind to enter the room. Within seconds, the flames leapt from the trash can to others. Within a minute, the fire was licking the underside of the stairwell. The school had no sprinkler system, so there was nothing to stop the hungry and growing expanse of flames unless someone happened to spot them. But with no one monitoring the boiler room, the fire was unleashed and soon out of control.

Climbing up the stairwell from the boiler room, the flames first licked the doors to the downstairs hallway. The firewall kept them at bay. Up another twenty or so stairs was the open hallway of the second floor. Using the stairwell as both kindling and a chimney, the flames roared up toward the unsuspecting children who were studying in the six rooms on OLA's second floor. There was nothing to stop the fire and, for the moment, no one even noted its presence.

At 2:30, the janitor had returned from the rectory to check the furnace. When he got to the door to the boiler room, he smelled smoke. John Raymond, who had four children attending the school, glanced through the window and noticed an orange glow. Backing away, he paused to consider what he had just witnessed. Even though he had seen it with his own eyes, he took another look. "Yes," he breathed, "the school is on fire."

Raymond's first urge was to rush in and fight the blaze. Yet he sensed the task was much too big for one man to handle. Running out a side door, he glanced up at the building. There was no smoke or flames on the outside. He breathed a sigh of relief knowing that for the time being it was contained to just the boiler room. Surely, he thought, it would not move very quickly in the brick building. Still, he knew a garden hose wasn't sufficient. The janitor also realized that he was not allowed to pull the fire alarm, so he had to alert someone who was authorized to pull the alarm. Running to the priests' residence, he shouted for the housekeeper to call the fire department, then rushed on to find one of the priests. The housekeeper must have not fully understood the excited Raymond, because it would take her another five minutes to make that call. Then she reported the address where she was working, a half block away and on a corner, rather than the school's address. This delay and confusion would indeed be costly.

As the janitor was looking for one of the school's administrators, several children, using the bathrooms on the first floor, walked right by the stairwell. The school's loud boilers drowned out the sounds of exploding glass and wood burning, and the heat was being carried up to the second floor, so none of the children noted either the heat or the noise of the fire. Even if the kids had spotted the signs of the fire, none of them could have reached the fire alarms. They were placed too high for the children to reach. Besides, they had been told to never touch them under any circumstances. If they were caught pulling an alarm, they knew that the nuns would deliver punishment.

While John Raymond was trying to alert those at the school, a salesman for a glue company was rounding the corner on Avers and Iowa when he saw smoke. Elmer Barkhaus stopped his car and studied the back of Our Lady of the Angels school. Smoke poured out of a door. Parking his car, he raced into a nearby store and alerted the owner, Barbara Glowacki. She seemed to ignore him, so for several minutes Barkhaus knocked on doors and rang bells until someone finally placed a call to the fire department. As the salesman raced up the streets, Glowacki, whose daughter went to OLA, walked outside. Seeing the smoke for herself, she ran back in and also called the fire department. Still, even as several urgent pleas were made to city fire officials, the schoolchildren and nuns were unaware of the nightmare that was quickly consuming their sanctuary. They continued classes just like they did every day. If the janitor had ignored the rules and regulations and pulled the alarm, then the children would have been out of the building by now. This haunted Raymond for the rest of his life.

A lay teacher named Pearl Tirstano was probably the first in the building to spot the fire. A student had opened one of the room's two hallway doors and Tirstano saw smoke rolling along the floor. It was wispy and thin, nothing to cause extreme alarm, so she ordered her children back into their seats and walked through the rolling smoke to the next classroom. Tirstano asked the lay teacher in the next room, Dorothy Coughlan, what they should do. An OLA rule stated that teachers couldn't allow the children to leave the building unless the fire alarm had been sounded. The only one allowed to sound the alarm was the Mother Superior. The two decided to keep the children in the room while Coughlan looked for the Mother Superior. When Coughlan couldn't find the head of the school in her office, she returned to her classroom and did the unthinkable. She not only broke the rules by ushering her own class from the building, she ordered Tirstano to do the same. Whether they were in too great a panic or perhaps fearful because they had violated a strict order, neither of the two women notified any of the other teachers.

A nun in room 208 watched as some of her students fanned themselves. The boiler must be acting up, she thought. It was hot in the room and growing warmer by the minute. She asked for a couple of her students to open some of the windows. She then went on with her teaching, unaware that two other teachers had already evacuated their classes because of smoke in the hallways.

Other classes on the second floor were also becoming warm. Finally, at 2:41, one of the nuns, Sister Canice, decided to notify the office about the problem. When she opened the door to the hallway, she was greeted by a wall of black smoke. Racing to the other end of the room, she jerked open the back door and was greeted not only by smoke but also sudden and excessive heat. "Dear God," the sister whispered. If it was a fire, she knew that she and her almost fifty children were trapped. Yet she also considered that it might not be anything more than the boiler acting up. It had acted up before. It was probably nothing to worry about. She then reassured her children and sat back down at her desk.

Outside, the school's neighbors, many of whom had children at the school, had first smelled smoke, and could now see it pouring from the stairwell. Flames were shooting from the door and some windows as well. They all wondered why the children had not run out into the yard. Why were they still in the school?

Inside smoke was billowing down the hallways and the temperatures were climbing. The wooden stairs had burned through from the bottom to the top and a few broken windows and the open doors on the first floor allowed oxygen in to feed the fire. Setting its sights on walls covered with dozens of coats of flammable paint, the hungry monster grew with each passing second. Its crackling could now be heard in every classroom and the shattering of walls and glass were even rocking the streets. Black smoke filled the sky and even those outside of OLA began to cough and wheeze.

Sounding like a bomb, the fire fully exploded into the second-floor hallway. The long, wide wooden-floored passageway was no longer just filled with smoke. Fire raced along the boards at a rate of several feet a second. With smoke pouring under the doors and through the transoms, the bright day suddenly became very dark for children in rooms 208, 209, 210, 211 and 212. They could no longer see their teachers and were having problems catching their breath. Though the nuns shouted for them to stay put, that help was surely on the way, the children were now more scared of the fire than the nuns. In mass they broke toward the windows, but for many of the younger children, the almost four-feet-high window ledges were just out of reach.

In room 209, fifty-five students were looking toward a veteran nun, Sister Davidis, to lead them out of harm's way. Her math lesson was forgotten as the teacher did her best with the limited resources she had. She ordered some of the students to close the transoms and block the cracks in the doors with books. Then she asked the students to walk over to the windows and call for help. What she couldn't know was that help would not get to her and the children in time to save them all. The fire department had just heard the news and the trucks were still minutes away.

If the black smoke was not problem enough, the classroom floors were also growing hot. Fire had now reached the crawl spaces between the floors and was even licking the transom windows.

Door frames and walls were in flames. Children were coughing, some crying, and the nuns could no longer see all of their charges. As the alarm finally sounded throughout the school, it appeared as if all the children left on the second floor were doomed.

"We need a ladder," a passerby shouted as she observed children's faces peering from second-floor windows. From the gathering throng, several men raced to their nearby homes to get any ladders they could find. Within minutes three or four wooden ladders were leaned up against the school, but none came close to reaching the second floor more than twenty-five feet above.

At 2:45, three minutes after the alarm sounded, children from the first floor were exiting the building. Father Joseph Ognibene, coming back from duties away from the church, spotted the fire. He managed to race up the building's smoldering fire escape and pull several children down the burning steps. When fire closed that route, the young priest climbed out onto a roof and noted another man, Sam Tortorice, literally pulling children through a window a few feet above. The two men formed a human chain and began passing children down to other volunteers manning one of the short ladders. As the moments passed, both men wondered where the firemen were.

As their heroic efforts continued, Ognibene and Tortorice could see flames crawling through the rooms and onto children's clothing. The fire was even beginning to shoot up the window frames. Many children were still trapped inside and their time was running out. The two rescuers worked hard, swinging kids one after another even as the heat pushed against them. Suddenly, without warning, the fire consumed the room, driving both of the men back. Neither of them knew how many children were instantly turned to ashes just a few feet from their reaching hands. Dropping to the ground, they fell to their knees. Then propelled by the screams of other children trapped in other rooms, they somehow found the strength to get up.

Sister Seraphica was teaching in room 210 when one of her students reported that the building was on fire. At first she thought the fourth-grader was just overexcited. But when she saw smoke coming under her door, she realized that there was real trouble just outside the room. Sensing she could not get away through the hall, the woman reassured her pupils by reciting a familiar prayer. Most did not join her, but instead raced to the windows and cried out for help. They sensed they had little time and only one avenue of escape. Within moments smoke was billowing into the room and the ten-year-old pupils were hacking and coughing, their eyes watering and the heat blistering their skin. When the doors blew open and the fire leapt into the room, panic took over.

Those who were tall enough and strong enough climbed into the windows. Some even used their classmates as ladders. The floor was now on fire and shoes were melting with feet still in them. Boys and girls were screaming as their hair glowed, flames crawling up their backs. Those lucky enough to be in the windows looked down at the concrete playground twenty-five feet below. They knew that to launch themselves into space and hit the ground would mean injury and maybe death. But it had to be better than burning to death. In a horrific moment of decision, those who had managed to get into the windows jumped. The sickening thud as they hit the playground slab was drowned out by the commotion around them. Yet their screams as clothes burned on their bodies and broken bones pushed through their skin were still heard by the many adults who had run to the scene. Strangers picked up bleeding children and carried them across the street to the wall of a candy store. They couldn't stay with them, though, because there were now scores of others jumping from the windows of Our Lady of the Angels.

In Room 212, the soot and smoke were so bad that Sister Therese couldn't offer the children any comfort. The nun realized that the only hope they had was to get to the windows and pray that the fire department's ladders would get there in time to rescue them. Heat was shattering the windows and light fixtures, and glass and ashes were pelting children as they ran to the windows. Some were trampled in the confusion. Then, less than two or three minutes after the alarm was sounded, the fire blew into the room like a tornado. Those who were closest to the windows jumped. Those scared of heights or traumatized by the turn of events simply stood rocklike as the smoke surrounded them and the flames grew closer.

Ronnie Sarno, a ten-year-old, was one of those who decided that jumping to a concrete playground offered a better chance of life than fighting the flames. Standing in a window, Ronnie saw his nine-year-old sister looking from a window in the room beside his. He walked the ledge to get to Joanne. "Come on," he begged. "We have to jump!" Joanne, like many other children, shook her head in terror. "No!" she screamed. She then added, "Ron, don't jump! Don't leave me!" As Ron slipped from his perch, his sister disappeared into the fire and the boy fell to the only safety he could find. He was joined by a few others, but not nearly enough.

The back stairwell would have offered some hope for those on the south side of the building's wing, but it was locked. The thirty or so students who were stuck in 207 had no place to go. John Raymond and a parish priest, Father Hund, fought their way through flames to that door, unlocked it and raced up the stairs. As the fifth-grade students screamed for help, the two men reached the room and led the teacher and most of the children to safety. They could not get to any of the other rooms, because the hall was now consumed by fire. As they pushed their way from the building, they heard the sirens that signaled the fire trucks were arriving. The noise was coming from the direction of the church, though, not the school. Raymond and Hund knew that the firemen were unloading at the wrong address. Somebody had to get to them before they unloaded too much of their equipment—children's lives were hanging in the balance.

The 85th firefighters had been given the church address by the housekeeper. When they arrived they saw smoke coming from almost a block away. Precious moments were lost. Children were burning to death and the first fire crew to answer the call was not in the right position. A fireman, scouting the scene, ran back to the truck and ordered them to move. When they finally got restarted and fought through now bumper-to-bumper traffic to the street in front of the school, the experienced firefighters were greeted by the sight of children launching themselves out of windows through the air and onto the concrete playground. They also saw faces, trapped in classrooms, peering down between clouds of billowing smoke and flames, too scared to even cry out. These children needed help, and they needed it now. The firemen called for backup as they tried to assess the situation.

Initially, the team tried to combat the fire, pouring water on the flames they saw. But the heat was simply too great for them to get close enough to make any headway. Trying to douse the fire enough to climb steps and work their way through hallways was simply not possible. The school's old tar roof, with its layers of material from several roofings, was keeping the flames inside the building. Rather than spread upward, the fire was now fanning throughout the structure. Realizing they couldn't contain the fire to save the children, rescuing the students became their primary push. With the clock ticking, civilians and professionals now worked together in an attempt to save children who were literally burning before their eyes.

Pulling ladders from the trucks, the firemen discovered, as had the neighbors, that their ladders didn't reach to the second-floor windows. The school's gate, locked securely, made access by ladder trucks that could reach the windows impossible. It seemed that every possible element was working against them.

Using ladders to knock the gate down, men rushed into the courtyard-like annex. They urged children not to jump as they pushed their ladders up to the brick walls. Yet ten-year-old kids whose clothes were on fire could wait no longer. They jumped time and time again, human fireballs, screaming in pain as they sought safety on hard concrete.

A couple of firemen with nets appeared. Setting up below a window, they begged for the children to jump one at a time. These children were burning alive, they couldn't wait. Those who stepped back into the room were turning to ashes in a matter of seconds. So they jumped together, some colliding in air, others missing the nets altogether, and a few finding a degree of safety in the firemen's arms. Nothing the firefighters were trying was working. It was mass chaos. With time running out for those in rooms 208, 210, 211 and 212, the men dropped their ladders and nets and struggled to catch children as if they were errant footballs.

One bull-like fireman, Charlie Kamin, got on a ladder and fought his way through heat and flames to room 211. Though the ladder was too short to safely use, he stood on the top rung and reached into the room. He had no time to ferry kids up and down to safety. Working with a wild abandon, he pulled kids from out of the smoke. The only ones he could manage to hold on to were boys; their belt loops offered something to grab. Many of the children he jerked through the windows were already aflame. Some had passed out. He yanked and dropped them one by one until the heat literally blew him off the building. The fireball that knocked Kamin from his perch burned the hair from his face. Even as he fell to the ground, he continued to reach for one last child. Thanks to the fireman's incredible courage and strength, ten lived who would have perished.

On the ground, as more trucks arrived, men were vomiting, not as much from the smoke as from the ghastly sight of helpless children being consumed by flames. Even those who had been through World War II had never seen anything like this. The smell of burned human flesh was everywhere. It was as if hell itself had enveloped the Catholic school. Men and women who knew no one who attended the school were hit with fits of tears. It seemed that every child that jumped became one of their own and every child who burned in a window was their baby.

As Kamin and others scrambled for new perches, the flames grew larger. With scores of firemen and hundreds of civilians now on the scene, the manpower was finally in place. But less than fifteen minutes after the call had been made, it seemed that no headway had really been made. The fire was still in charge and mortals seemed helpless against it. Though they knew children were still alive in the building, their screams could still be heard, it was getting harder and harder to reach them. In parts of the school the fire was now burning at more than fifteen hundred degrees. Rescuers on ladders witnessed children ignite in a terrifying instant. Some turned into candles even as they got ready to jump from a window.

"My baby is in there," mothers and fathers cried out. They grabbed firemen and priests and urged them to save the children. Yet even the professionals had no answers. In the past ten minutes incredible life-saving efforts had been made. Some children had been saved; they were lined up against walls and lying in the backs of cars. But so many seemed to still be in the building.

By 2:55, just ten minutes after the first crew had arrived on the scene, the second-floor roof and ceiling came crashing down. Unable to burn through it, the fire had simply consumed all the supports that held the roof in place. When the tar and timbers tumbled down, it took the children in room 208 with it. They were alive, firemen were within reach, then suddenly they were gone. It happened that quickly.

By now all of Chicago knew about the fire. Smoke could be seen rising from almost all parts of the city. Radio and television reported it. Ten ambulances were on the scene, scores of firetrucks and police cars had arrived. The people from the neighborhood around OLA were there, too. Healthy children, those who had made it out of the school, either watched in horror or broke ranks and ran home. Parents arrived looking for their own kids. With nine hospitals receiving injured students and teachers, no one knew who was dead and who was alive. No one knew how many were left inside the burning structure. A priest told the families and the press that he thought as many as a dozen children and a couple of teachers had died. Considering the magnitude of the flash fire, that number seemed to indicate that OLA had been fortunate. Yet firemen like Charlie Kamin knew better. He had seen more than a dozen children left in the one room where he had been feverishly trying to pull them to safety. He knew the final body count was going to be very high.

Within an hour of the first call to the fire department, the flames were under control, Chicago's mayor Richard Daley had arrived and the search for bodies began. Many were found in piles, literally melted together. A few were still sitting at their desks or perched next to a window. In room 212, twenty-seven still sat waiting for a rescue that never came. In room 210, Sister Seraphica was found, her thirty charges literally wrapped in her dead arms. In room 208, Sister Canice died too, along with ten children who had looked to her for help and guidance. Across the hall in 211, fireman found twenty-five more little bodies.

Newspaper reporters who walked through the school with the firemen were sickened by the scene. Many of the children had been burned to the bone. Flesh had been melted by the extreme heat. What was left of their faces reflected the pain that must have consumed the children in their last moments. Some bodies had turned to ashes and dust. And outside, scores of parents kept asking, "Have you seen my Johnny or Mary or Sue?" Some of those inquiring were firemen who had given their all not only for some nameless child, but for their own too.

Firemen would be haunted by the tragedy for years. Many even quit the force after the fire. The images they mentally recorded as they shifted through the rubble—the burned coats, shoes, books and chalkboards still filled with an unfinished schoolday's work—were too much for them. In one classroom a firemen broke down in tears as he discovered an art project signed Joseph. With great care the boy had fashioned a plaque that read, "I Joseph, promise to do my best, to do my duty to God and my country, and to be square, and to"—he was almost finished when the fire had hit the school. The flames took the child, burned him to ashes, but left his fragile unfinished work—a Christmas present he had planned on giving to his parents—intact. That and so much more was simply too much for the firemen to bear.

That evening, as the smoke lingered in the air, husbands and wives split up and went from hospital to hospital. If they didn't find their missing child or children in any of these facilities, they worked their way back to the county morgue. There, among the priests assigned to give last rites, the frightened parents prayed as they carefully looked at each of the day's victims. Eighty-seven children and three nuns lay silently under white sheets. For many, only dental records revealed their identities.

For the next several months, as scores of badly burned victims lingered in hospitals, a few succumbing to their injuries, Chicago officials tried to figure out what had happened. Unofficially, most felt that an unknown student at the school had either accidentally set the fire by tossing a cigarette into a wastebasket or that a student, who two years later would be arrested for multiple arson attempts, had deliberately set the fire. Ultimately it didn't matter. If the city had forced older schools like Our Lady of the Angels to renovate, install sprinkler systems, put in firewalls and fire doors and enforce strict codes on how many students could be housed in each room, then the disaster would have surely been averted. Yet even today, in the midst of the technological age, many ancient schoolhouses are no better protected from fire than was Our Lady of the Angels in 1958.

Christmas 1958 came and went. OLA made plans to rebuild and the families whose lives were so deeply scarred by the fire tried to regroup. Yet even as winter gave way to spring and spring to summer, there seemed to be constant reminders of what happened that day. Children, teachers and parents had constant nightmares. In some minds, the fire was still burning. If the flames ever did really die, they did so on August 9, 1959. On that hot summer day, William Edington passed away. William had survived the fire and two dozen operations. His spirit was willing, but his heart finally gave out. The final death total was now ninety-five.

# 9

## Johnstown

### Rich Man's Flood, Poor Man's Disaster

## 1889

In an already wet month, beginning just after midnight on May 30, 1889, the skies opened up and poured out torrents throughout western Pennsylvania. In areas along the Conemaugh River as much as eight inches of rain would pelt the ground that day, swelling rivers over their banks and sending water into streets and homes. It made for a soggy Veterans Day celebration; nevertheless, Johnstown's thirty thousand residents were determined to honor those who had served the Union in the recent Civil War.

The War for the Union, as it was most often called in this part of the United States, was a very recent memory for many of the local townspeople. Scores of the city's best-known men had worn blue between 1861 and 1865, several of them even winning medals and earning the title of hero. On this dark day, many of these now aging soldiers sucked in their bellies and tried to fit into their uniforms one more time. In spite of the rain, the men gathered and marched up Main Street and over to Adams and out to the Sunnyvale cemetery to acknowledge those who had not survived the conflict. They then paraded back downtown to the local theater, where they were guests of honor for a musical production. Except for the rain that continued to fall in buckets, nothing seemed unusual.

The next morning, the holiday over, most men got up early and headed off to work. It was Friday and just another ten-hour shift in a grueling six-day workweek. As the men walked to work in the steel mills and rail yards or hurried to the businesses that supported these industries, they grumbled about the water that flowed down the streets. It was ankle deep and rising with the rain still falling. It was typical of this time of year. High water in May and June was as normal as sunshine was in August and September.

Downtown Johnstown had the misfortune to have been laid out at the juncture of the Conemaugh and Stonycreek Rivers. When it rained hard, water from both of these flowed through not only the streets but many of the downtown buildings as well. Ten other smaller communities that were built along the river in the deep valley above Johnstown faired no better during the rainy season.

By ten o'clock, Johnstown businesses such as the Cambria Iron Works had given up. With six inches of water flowing through their plants, rather than try to keep the lines moving they simply sent workers home. Even though they had the day off, most of these men knew they would spend a good part of it working with their families, moving rugs, furniture and other possessions to higher ground, usually the second floor or the attic. Yet in spite of the rising water and the continual rain, few saw this as anything but a temporary problem.

Fourteen miles to the north, there was much more concern. The South Fork Dam was holding back more than twenty million tons of water. The dam was part of a failed canal project that was supposed to connect Philadelphia to Pittsburgh. A lake had been constructed to supply the water for the canal. However by the time the earthen dam was completed a decade before the Civil War, rail service had made the canal an outdated idea. Worse yet, no one wanted to buy the dam or the lake from those who had constructed it. So, beginning in 1854, before it was put into service, the dam began to deteriorate. In 1862, it finally broke and a fifty-foot section of the two-hundred-foot-high dam washed away. The Pennsylvania Railroad, the dam's owner, couldn't have cared less. They didn't bother fixing it, and over the next fifteen years they pretty much forgot about the property.

In 1879, Benjamin Ruff discovered the broken dam and dry lake bed and decided that he had found a gold mine. He envisioned a picturesque lake surrounded by huge trees and beautiful mountain views. He saw unlimited wildlife, rugged trails and unspoiled wilderness. Rather than a wasteland, the developer viewed the area as the ultimate vacation retreat. Ruff was sure that he knew a lot of very rich men who would pay a great deal to claim this part of Pennsylvania as their own.

Benjamin Ruff drew up plans for his dream resort. He then managed to sell that plan to his wealthy friends. Within months he had sold more than a hundred annual memberships at two thousand dollars apiece. This was a huge sum in the late 1800s. With plans such as fixing the dam, building clubhouses and cabins and stocking the new lake with prize game fish, the South Fork Fishing Club soon boasted such famous names as Mellon, Carnegie, Frick, Knox, and Harper. The richest men in Pennsylvania wrote checks and watched as the hole in the dam was filled with rocks, dirt and refuse, and the entire structure was extended to almost double its height and width. In short order the lake filled and the South Fork Fishing Club quickly became one of the most exclusive resorts on the East Coast. It was beautiful, unspoiled and, best of all, the rich would not have to mingle with anyone outside of their social circles.

Those below the dam watched from a distance as Conemaugh Lake grew to more than three miles in length. Many were concerned that the old dam, which had ruptured once when it was almost new, would not be able to hold back that much water. In some places along the dam's earthen wall, the water's depth was now almost five hundred feet. Worse yet, the original broken drainpipes had not been repaired, only plugged, and the old spillways were locked in order to keep the newly stocked game fish from escaping downstream. Even though engineers from the iron companies pointed out to the wealthy vacationers that their dam was not sound and spillway and drain relief was needed, no one listened. The millionaires were too busy building cottages, installing indoor plumbing and staging regattas to worry about a dam that appeared fine to them. As a decade passed and the South Fork Dam stood strong against the weight of the lake, many living below the structure in the valley forgot about it altogether. On rainy days when a few excited men would voice concerns, they were generally laughed into silence. After all, it was argued, Mr. Ruff has stated many times in town meetings that "you and your people are in no danger from our enterprise."

What Ruff had dismissed and the common people in the valley did not know was that the South Fork Dam could not survive if water ever spilled over its top. With no solid core, the earthen dam would then quickly turn to mud and crumble. With the spillways blocked and the drainpipes not functioning, there was no way to drain water around the dam. Thus, if enough water did fill the lake, engineers knew that not only would the dam collapse, but the water that flowed through the hole would pour down on the valley like Niagara Falls. On the morning of May 31, 1889, there were four and a half billion gallons of water behind the South Fork Dam, more than had ever pushed against it, and the pressure that water was exerting on the dam was showing.

The South Fork Fishing Club was basically deserted. The millionaires and their families would not start arriving at the lake for another two to three weeks. Yet three men were standing along the dam that rainy morning, each watching as the old edifice began to leak. John Parke, an engineer in his early twenties, had long felt that the dam was not sound. He had even tried to tell the membership that on several occasions, but his views were quickly dismissed. As he watched water pressure tunnel through sixty feet of dirt and blow holes in the creek side of the wall, he shook his head. From those breaches, streams of water shot out more than fifty feet. The dam looked like a showerhead running at full pressure.

Beside Parke, E. J. Unger, the club's president, and W. Y. Boyer, the superintendent, watched with interest but not as much concern as the engineer. They still felt the dam would hold. Besides, one of them pointed out, the rain was slacking off. When it quit, they argued, the danger would pass.

Sixteen miles and almost four hundred feet below the dam, the residents of Johnstown were simply trying to stay dry and, like the men at the club, hoped the rain would soon end. None of them were worried about the dam; after all, most believed that the richest men in the world would have already done something if the South Fork really was unsafe. Few of them understood that the richest men in the world didn't view the dam's problems as their concern and thought less of those who lived in harm's way below the structure than they did the fish that swam in the lake.

By 11:00 A.M., most in town were trying to find a way to prepare the noon meal. Because many kitchens had several inches of water flowing through them, the task was more difficult than usual. For most people, it appeared that dinner, as the noon meal was called then, would be cold leftovers. As housewives began to put meals together, John Parke was standing to the side of the South Fork Dam watching the water lap along the top. An hour before he had ordered a crew that had been digging sewer lines to try to cut a channel around the side of the dam. He now knew that it was too late. Parke had no doubt that the dam was going to break. Rather than allow countless men, women and children to die, he jumped on his horse and headed toward Johnstown. At every farmhouse he stopped and shouted his warning, "Head for high ground, the South Fork Dam is about to break." He then raced to the South Fork railroad tower and yelled the same warning to the telegraph operator. Feeling assured his warning would get to the thousands who needed to hear it, Parke rode back to the dam, not knowing that the telegraph lines from South Fork to Johnstown had been washed out by a rock slide. He also couldn't have guessed that most of those who had heard his message had simply dismissed it and continued what they were doing.

Parke was ignored for two reasons. The first was his age. Men in their twenties were still considered children in many people's minds. The judgment of the young was simply not credible. The second reason few noted his warnings was that they believed he was crying wolf. Each of them had been told many times before that the dam was going to break, and on those occasions they had bundled up their families and climbed the steep hills above the creek only to feel foolish when the floodwaters never came. So hundreds who could have been saved by Parke's warning now had only hours to live.

When Parke returned to the dam, he was horrified to note that holes as large as whiskey barrels had now opened up in the wall. Slowly the holes enlarged and by three in the afternoon water was pouring through scores of them. Ten minutes later, a section the size of a football field crumbled and slid in one massive sheet into the ravine. As it did, a forty-foot wall of water leaped from the lake and jumped into the creek. The water thundered like a thousand cannons as it poured from its home into virgin territory. Except for the three men beside the dam, no one heard the explosion of water and mud.

With the force of thousands of locomotives, Lake Conemaugh shoved away all the soil, rock and lumber that had once held it in place. In a matter of an hour, the lake's entire water supply would pour through that huge gaping hole and let loose millions of tons of liquid death on the valley below. Even though they should have been ready, even though some were prepared, even though engineers had predicted the dam's demise for years, few below the dam felt any concerns.

Two farm families had heeded Parke's warning and headed for the hills. Even though they had a ten-minute head start, they had only climbed high enough to barely escape the water's reaching hands of death as the flood came rushing past. They looked down from a hilltop as everything they owned was swept under the waves and their houses and barns were turned into boats. In the blink of an eye, years of hard work and dreams simply vanished. The dozen mute witnesses realized what was ahead, but they were powerless to do anything about it.

The first small town in the torrent's path was South Fork. Because of the high water caused by earlier flooding, the low-lying village had been evacuated. Though no one was there to witness the horror and no one lost their lives when the water hit, less than eight minutes after the dam broke the town was erased from the face of the earth.

Just outside of South Fork, a train bound for New York was waiting on the tracks for a mud slide to be cleared. The engineer just happened to be looking upriver when he noted the huge wave of water and debris surging toward him. Jerking the engine into reverse, he guided the train across a bridge over the creek as the water bore down on the trestle at thirty miles an hour. With terrified passengers staring out windows, the passenger train lurched along, the water growing closer by the second. The last car barely cleared the bridge when the wave hit, smashing the structure and a freight train on the other side into pieces. As passengers caught their collective breath, the water raged through the creek bottom and entered the channel where South Fork Creek joined the Conemaugh River. Two miles downstream, less than four minutes ahead, lay the tiny town of Mineral Point and eight hundred unsuspecting people.

A telegraph operator a mile above the town spotted the wall of water. By now it was black with debris, spraying mist hundreds of feet into the air and roaring like a cyclone. Though his lines were down, the operator shouted a warning to train engineer John Hess, who was standing below the telegraph tower. Hess jumped in a lone engine, tied his whistle down and shoved the locomotive into reverse. With the whistle blasting a warning, he pushed down the valley toward Johnstown while the telegraph operator leaped from his platform and ran for the hills.

Mineral Point was ripe for the picking. Most of the town rested on the flat bottom along the river. Few homes were built on the hillside, so most residents had no real chance to escape. Even as Hess's whistle blew, it was already too late. The water was too close and moving too fast. Hundreds of confused people bolted from homes and stores when they heard the whistle. Standing in the streets and yards, they stared horrified at the wall of water. Within seconds they were racing toward high ground, but only a few with speed made it to safety. Almost everyone in town suddenly found themselves, as well as everything they owned, covered with cold water. A few were hit by the wall and crushed, a few drowned, but most simply clung to pieces of houses and barns and tried to ride the makeshift boats to safety.

Scores of men who had been working on farms along the hills raced toward bridges that had been built high over the river. Realizing that the forty-foot wall of water would be unable to reach the bridges, the men lowered ropes into the water to try to rescue their friends from Mineral Point. Scores of people grabbed onto the ropes and were pulled to safety. These acts of selfless courage would continue for hours and hundreds would ultimately owe their futures to men who put their lives at risk to pluck victims from the roaring water.

The angry wall of water destroyed the buildings in Mineral Point as easily as a child pushed away marbles. In less than a heartbeat, it swept the streets clean. Next came the towns of East Conemaugh, Woodvale and Franklin. By now the waters had slowed a bit. The flood's surging tide was filled with hundreds of thousands of tons of debris, including homes, barns, livestock, train cars, trees and mud. This gave a few people in the tiny hamlets below a chance to spot the wall of water in time to get to higher ground. Still, many were caught in their homes and taken to their deaths with no warning at all.

John Hess and his locomotive had outraced the flood to the Conemaugh Yards before he abandoned his engine and jumped out. As he ran uphill, his tied-down whistle continued to alert those as far away as Johnstown that something was amiss. That whistle saved the lives of hundreds, but thousands of others never realized why the locomotive was making such a mournful racket.

In the train yards that day sat an express train. Its passengers, many from Philadelphia, were anxious to get on with their trips. Yet the railroad didn't seem to want to hurry. The flood had played havoc with their rails the last few weeks and the Pennsylvania Railroad officials wanted to make sure that the track was safe. As the dispatcher waited for the go-ahead from the inspectors, passengers grumbled in the cars. It didn't take long for Hess's whistle to get on their nerves. Two twenty-year-old women, Jennie Paulson and Elizabeth Bryan, outfitted in their spring best, wanted the noise to stop. Like most passengers on the express, they didn't comprehend that the whistle meant trouble. As they chatted with the girls, newlyweds Charles and Edith Richwood also wondered what was the matter. They questioned a confused porter as to when they would be able to continue their trip to New York. Others on the train spoke of the whistle and their disgust with the delay, but no one moved. Then someone observed a huge wall of dark water rolling slowly toward them. It didn't look real and the few who spotted it wondered what it was. Mesmerized by the swirling movement and frozen by the distant roar, the fifty or so passengers sat silent. Finally, someone realized that a massive flood was heading their way. The cars erupted in panic.

Henry Smith, the general manager of the Associated Press news service, was one of those who finally figured out what was bearing down on the train. He would later describe the forward surge of the flood as "the black head of a monster." He watched as frightened passengers pushed past each other, leaping from the train and running as fast as they could up a slippery hill. Their goal was to reach a high point a hundred yards away. Many made it, but some delayed too long.

The two young women who had worn their best for the train ride decided to put on their rubber boots before exiting the car. Their boots never touched the ground. Paulson and Bryan would die as the wall of water rolled their car like a ball through the rail yard.

The newlyweds simply didn't run fast enough; the Richwoods

would be caught by the water and swept away toward Johnstown. As the couple fought to find something to climb onto, they watched fifty-ton locomotives swept off the tracks and tossed like toys ahead of the waves. All around them, houses bobbed like fishing corks. What had seemed like the beginning of a beautiful life together now appeared to be a single moment of bliss followed by immediate doom. Locked in a lovers' embrace, the Richwoods fought the waves and clung to each other.

Just a few feet ahead of the newlyweds, Henry Smith turned after he hurdled a ditch filled with surging water and looked again at the wall of water. He saw people inside the wall, their mouths locked in screams. The AP head then glanced back at the passengers who were behind him. Some had quit running and were simply staring into the teeth of the immense expanse of water and debris. Some were screaming, but Smith couldn't hear them over the roar created by millions of gallons of racing water. Terrified, Smith ran on, turning again only when he felt he was high enough up the hill to be safe. He then looked down at the rail yards just in time to see the wave bury the passenger train and smash into twenty parked locomotives. The huge black engines were tossed ahead of the wall as if they were sticks and pushed like huge bowling balls through houses before being enveloped in a sea of water and mud. Riding the crest just ahead of those locomotives were the Richwoods. Smith said a prayer for a couple he knew was doomed.

Hundreds were now fighting for their lives in the swirling yellow cauldron, surrounded not only by lifeless debris but more animals and people than they could count. Bodies suddenly appeared everywhere, having been swept off trains, out of stores and homes and even grabbed from trees. Unless they had made their way to the hilltops, no one was safe. Not even brick buildings could stand up against the tons of water that had torn the dam apart. The only chance offered to many was that the water was now moving in fits and spurts as the sheer volume of wreckage it was sweeping in front of its crest created a series of dams. Yet just like the one at South Fork, none of these held either.

At 4:10 P.M. the giant wave finally touched the residential streets of Johnstown. Its timing could not have been worse. The rain had let up, the water in the streets had begun to recede, and people believed the worst was over. On the crest of the hills, where thousands from the upper valley had fled, witnesses looked at the wall of water pushing the debris field along ahead of it and then to the city, its citizens calmly going about their Friday-afternoon business. They had to wonder why no one was running, why the streets and the hills just above town were practically deserted. Then, when the crashing water hit the first homes in town, Johnstown woke up. The *New York Sun* would later print this eyewitness account of that moment:

> In an instant the deserted streets became black with people running for their lives. An instant later the flood came and licked them up with one eager and ferocious lap. The whole city was one surging and whirling mass of water, which swept away house after house with a rapidity that even the eye could not follow.

A witness likened what he was watching to an avalanche. It was probably as good a description as any. As the wall of water rolled into Johnstown it contained more than 100,000 tons of rocks, locomotives, freight cars, wagons, iron rails, trees, homes, barns and wire. Added to this total was 16,000,000 tons of water moving at almost a mile every two minutes. It was a giant killer with an appetite that could not be fathomed. Worse yet, as soon as it reached the heart of town, the water rushing through the flooded Stonycreek River would feed it as well.

John Fenn, his wife and seven children lived on Locust Street. He was working at his hardware store when he was warned that a flood was about to strike the city. He made it to his front gate when the wall hit. His wife was watching from the second floor as the little man was pushed under the two-story wave. Mrs. Fenn felt her house give way at the same instant her husband went down. Gathering her brood around her, she made it to the attic and out on the roof. Within a minute that roof was all that was left of the Fenn home. As eight frightened faces looked out over the now pounding surf, they were greeted by the moans and screams of hundreds of others trying to ride out the flood. Fenn tightly held on to her children as the water smashed against their makeshift raft, but each time the flood tossed the section of roof against another pile of debris, one of her children was knocked into the raging current. Over the next five agonizing minutes, she lost all seven. Collapsing on her perch, she gave up on life as her last child, Francis, was sucked under the water. Now not caring if she lived or died, she waited for the flood to claim her. Yet miraculously, she was yanked from the water five minutes later. She would live with the horror of what she saw for decades, never completely accepting or getting over the loss of her family on that day.

The Fenns were hardly alone. Countless families were dashed by the water. The Hulbert House, a strong stone building, was built to withstand anything. Because it was the nicest and newest hotel in Johnstown, scores of families sought safety there as the wall approached. J. L. Smith had taken his whole family there before returning to wait out the flood in their frame home. Smith watched the wall attack both his house and the huge building. Surprisingly, his house of sticks stood against the tide but the downtown buildings caved in one by one until the wall of destruction arrived at the hotel. The Hulbert House withstood the millions of gallons of water and tons of debris no better than any of the older buildings. Husbands and wives, desperately clinging to their children, were instantly crushed when the walls gave way. Smith's entire family died in the rubble as he watched the building go down.

Bertha Caldwell had come home from Utah to visit her folks and, in spite of the weather, was enjoying her time in Johnstown. She was on the third floor of the Caldwell home when the wall of water hit. Caldwell at first thought she heard a train, the rumble sounded so much like coal cars chugging down a track, yet when the whole house shook, she knew it was something else. Racing to a window, she saw the wall of water spitting out homes and people and steadily marching toward her. With her parents and younger brother and sisters at her side, she prayed. Her prayer was answered, because even though the waters came three feet up into the third floor, the brick structure somehow stood firm as all the homes around it were swept into the waves. Yet as she looked into her father's eyes, Bertha could sense he believed it was only a matter of time before, as he whispered to her, "We join the Lord."

All around the Caldwells' home men, women and children were struggling to stay afloat in the torrent. Looking at the faces of people she had known all her life, Bertha felt helpless. Then, as the debris piled up and pushed the big house off its foundation, the woman realized that soon she would have to join those who were fighting for their lives in the water. Grabbing her sister, Bertha didn't wait for the house to cave in—she leapt onto a roof that had floated against their home. Her family followed. They climbed over debris and continued to float through town for an hour, but they were never sucked into the muddy pool of death. Unlike most caught in the middle of the flood, they survived.

A few homes would somehow stand against the flood. Reverend David Beale's house was one of those. From the attic of the Presbyterian parsonage, Beale did more than watch the calamity. Risking his life, he pulled more than twenty struggling victims to safety.

Henry Kock and his porter, George Skinner, also watched from the top floor of a building as the wall of water rolled in. Realizing that they needed to help those caught in the water, Skinner grabbed Kock by the ankles and literally tossed him out the window. Holding the smaller man, Skinner braced himself as Kock grabbed anyone who floated by. The two men saved more than a dozen people who otherwise would have surely drowned.

In the midst of the most terrible moments of the flood, scores like Beale, Kock and Skinner dangled by ropes, jumped from debris pile to debris pile, leaned out of windows and reached down from roofs to save those caught in the water. Hundreds were brought out of the water through these courageous acts. Yet for more than two thousand others, there would be no salvation.

The flood's destruction would roll through Johnstown at three blocks a minute before hitting the Pennsylvania Railroad Bridge. The mighty stone-and-steel structure finally managed to do what nothing else had. It withstood the barrage of tons of debris and water. It stood strong even as houses, wagons, train cars and people were dashed against it. While downtown Johnstown tumbled along like a series of snowballs rolling down a steep mountain, the bridge held firm. In doing so, thousands of structures, tons of trash, miles of telegraph lines, countless dead animals and hundreds of people, both living and dead, were stacked up against the bridge in a debris field that would reach almost as far as the eye could see and cover more than sixty acres. The water was forced to go over, under, around and through the wreckage and did so with a rage that few had imagined even in their worst nightmares.

Above the roar of the flood, the crash of falling timber, and the swirl of rushing waters came the groans of the dying, the wails of the mangled, and the agonizing cries of the trapped. When the cold water hit the furnaces in the steel mills they exploded like bombs and the earth shook as if being hit by an earthquake. Churches and businesses were mowed down and those inside were either drowned or crushed. Women holding dead children were mouthing unheard words to hundreds on the shore and men who had refused to grab ropes because they were trying to get to their own children were floundering hopelessly in the rapids. Many hearty souls on the hilltops turned away from the scene in shock. Others simply stared blankly as they watched men, women and children die in every possible way imaginable. These few minutes made the horrors of the Civil War seem like a child's backyard game. Here there was no mercy, no chance to lay down a weapon and surrender, no retreat and no route for escape. If you were in the swirling water, only God could pitch you the right way to either end your suffering or allow you to gain a foothold for rescue.

In the midst of the chaos, witnesses were gripped by an eerie sight. Two huge locomotives, steam still belching from their engines, were being hurled ahead of the wall as it passed Johnstown.

The engineers, knowing that their lives were now numbered in seconds, were blasting their whistles in an attempt to warn those downstream to head for the hills. Neither man stopped their blasting until the wall swallowed the engines.

As people struggled to find solid ground, the flood began to construct islands of hope. Stacking homes and railroad cars one on top of another, massive mountains of debris offered safety to scores who had been trapped in the water. Some had been dodging death now for almost thirty minutes. Swimming, thrashing and praying, they grabbed onto the mounds and pulled themselves up, then turned to reach for others. Soon, what used to be parts of houses and businesses were refuges to those who, just minutes before, had thought they had no hope. Included among these were the newlywed Richwoods who had barely escaped the express train.

Even as miraculous acts of heroism and luck were played out in the midst of the forty-foot wall of water, with one great swoop the wave now pushed over three thousand houses, hotels, stores and factories in less than ten minutes. Like long rows of dominos they were knocked to the ground and sent crashing, tumbling, and floating down the roaring torrent. This seething mass, speckled with people begging for help, was on a course to strike Cambria City. Unlike Johnstown, the next community would have some warning.

Western Union station manager Harriet Ogle, a heavy-set woman with the courage of ten, watched the wave come toward her. As she stared at the massive wall of destruction Ogle continued to pound on her telegraph keys, never giving thought to leaving her post. Even as the water approached her, even as she stared into the teeth of a wave that was spitting out bodies of people she knew, the woman pounded out the warning on her key to the cities below. Her last words, "This is my last message," became a legacy, preparing those downriver for what was crashing through Johnstown at that very moment. Because of Ogle's sense of duty, Cambria City, Coopersville, Morrelville, Sand Hallow and Sheridan Station would not suffer the great loss of life that had struck Johnstown.

Even as the wave passed, most of Johnstown was covered by

more than twenty feet of water. Thirty minutes after the first wall hit, people were still drowning in the swirling rapids that flowed through the city streets, and houses that had withstood the first pounding were now simply falling apart. The few that managed to get to sections of town where the water had receded were simply lost. There were no markers left. No street signs, no foundations, no bricks, no power poles, just mud and flat space. It was as if most of the business and residential districts had simply disappeared. Yet as bad as things seemed, the real agony was only beginning as the city clock struck five. After the unthinkable baptism by water, Johnstown was about to feel the wrath of fire.

Thousands had somehow lived through the flood only to find themselves trapped in the wreckage that had smashed against the railroad bridge. As they sat in their temporary jails, hundreds of these survivors voiced prayers of thanks. They had survived wild rides atop the forty-foot waves, they had dodged tons of wreckage and had managed to float to seeming safety. Then, as they waited for rescuers to find them and dig them out, they smelled smoke.

The flood that had buried the iron works and taken hundreds of rail cars for rides across town had also distributed coal embers throughout the wreckage. With oil, wood, paper, cloth and other combustible materials in ready supply, and gas leaks everywhere, the embers quickly created little fires that burned throughout the wreckage. As the flames spread and twilight began to settle over the community, those who were trapped became aware of the fate that now awaited them. If they couldn't get out, they were going to burn to death.

Although a steady rain continued to fall, the debris field soon became a furnace. With the firelight providing illumination, scores of frantic men jumped onto the mass of wreckage and started tearing it apart. They pulled screaming victims from the carnage even as the flames grew closer. Over the course of an hour, as the fire ate through the pile of debris, as gas triggered explosions, teams of fearless men raced ahead of it with axes and hammers. They ignored the dead and looked for the living. Whenever they heard a voice, they stopped and went to work. They continued to chop and slash until they managed to pull a person from the tangled mass or until the flames singed their hair and clothing and forced them on. It was a race against time, and even though they were surrounded by millions of gallons of water, they had no means to fight the flames. If they couldn't dig fast enough, then another person would die.

J. M. Frontheiser had already lost a wife and child to the flood. Now, as he combed the debris that had stacked up against the bridge, he heard familiar cries that brought him some comfort. Looking through the jumble of what had been parts of the valley he loved, he discovered a son and daughter tangled up in the debris, their heads barely rising above the water. Frontheiser desperately pulled away the twisted wires, boards and even dead animals from around his children. Yet as he worked a flash fire ignited to his left. With his children screaming, begging him to get them out before the flames could consume them, the man, now feeling the heat at his back, pulled at the wreckage and his children. He managed to pull them both free just before the fire erupted in the tangled mass of rubble that had held them in place.

As Frontheiser worked his way across the mass of wreckage toward the shore, he heard the cries of hundreds of others all around him. With the flames chasing him and his children in his arms, he couldn't stop to render aid. Yet he still could see the horror that was all around him. Glancing out into the water he observed a woman on a floating roof. She looked like a nightmare, her clothes a mess, her face covered with mud. As the current pushed her into a wall of flame, she stood and started to sing "Jesus, Lover of My Soul." The song and that image would haunt Frontheiser and his children for the rest of their lives.

The fires would kill more than three hundred people. Most would die slowly, always aware that they were so close to safety but simply couldn't get free in time. Many begged to be shot, some prayed, and when the fire finally caught them, their screams could be heard all up and down the valley.

The night brought unimagined darkness and horror. Those trapped by floodwaters had no means of escape, no food and no contact with those on the hillside. No one lit matches or lamps because of the leaking gas mains. The survivors were also scared that another wall of water would come suddenly and take them like it had so many others. They kept looking and listening for the next flood.

On that dark night four women gave birth. Each had a boy. Two of the new mothers named their offspring Moses, because, like the biblical hero, the boys had been brought out of water. The other new arrivals were, for obvious reasons, christened Flood.

As the newborns cried out for life, fourteen miles above, John Parke wondered what the wall of water had done to the city of Johnstown. His only comfort came from the fact that his warning should have spared a majority of the people in the towns and farms below. When Parke later found out that 2,200 died and that few heard his warnings because the telegraph line was down, he collapsed in sorrow and grief. "If only they had listened to me," he sobbed. His words were meant not for the citizens of Johnstown, Mineral Point and the other communities, but for the members of the South Fork Fishing Club who had hired him.

The next day stunned men and women looked at a valley that had been changed so dramatically by the collapse of the South Fork Dam. There was chaotic trauma on every corner of every town that had been hit, yet it was also strangely peaceful as a warm sun lit up the clear sky. Everywhere people turned there were bodies, some with hands sticking up out of the mud, others with faces frozen in silent screams. Orphans were crying out for mothers and fathers whose bodies would never be found. Shock so numbed people that many moved along the muddy ruins of streets in silence. Survivors were either wallowing on muddy ground or packed into homes that had been built high on the hills.

With no food or medicine, little housing and hardly any medical care, tens of thousands were in great need. One of the first to reach the scene was sixty-eight-year-old Clara Barton and a Red Cross team. America's most famous nurse and her organization would reach more than twenty-five thousand people with aid and administer more than three million dollars worth of services in the months that followed. Yet while Barton and her people, along with millions of Americans who were deeply troubled by the loss of life and property, struggled to deal with the aftermath, another group was examining an entirely different problem that seemed pale in comparison.

Many of the rich sportsmen of the South Fork Fishing Club seemed more concerned about the loss of their game fish and boats than they did about the enormous losses suffered in the towns below. While tallying their losses, they also refused to accept any blame. When people pointed fingers of guilt at them, they simply pushed the matter over to their lawyers. At first the lawyers claimed that the flood had been caused by another failed dam. Then, when that lie was exposed, the attorneys stated that the businessmen could not be held accountable for a dam that had been built long before they had become a part of the club. It didn't matter that the South Fork Club members had the resources to fix the structure but chose not to.

Newspapers across the nation did not buy these claims of innocence. The *Harrisburg Telegraph* stated, "50,000 lives in Pennsylvania were jeopardized for eight years so that a club of rich pleasure seekers might fish and sail and revel in luxurious ease during the hot months."

A coroner's inquest also held the club responsible for "gross, if not criminal, negligence and carelessness in making repairs from time to time." The accusations continued for weeks, but to no real avail. Although it was money not spent that had created the problem, money well spent gave the members the power to ward off the concerns of the communities and companies in the valley. The club and its members were not held responsible even once and did not pay out a single dime in damages. Eventually they assumed the role of victims. After all, they argued, their club had been ruined just like much of the valley. Of course, they had insurance, while most in Johnstown and the other villages did not.

In the tragedy 311 children lost their fathers, 156 their mothers, and almost 100 lost both parents. The area's widowed women numbered 125, and 200 men lost their wives. A full 99 families were wiped off the face of the earth. Yet the South Fork Fishing Club didn't care enough to offer relief to those who had suffered this greatest loss of all.

Johnstown and the valley lost all their railroad tracks, a large portion of their doctors, all major businesses, schools and churches, jails and police force. The gas and electric utilities were destroyed, as was all telegraph service. Every mill, factory and train station lay in ruins. There was no clothing, no food, no medical supplies or blankets. Everything had gone downriver. It would be weeks before even the town's wealthiest survivors had a change of clothing or enough food for more than one meal a day.

Johnstown slowly rebuilt, but is smaller today than it was 110 years ago. The South Fork Dam was never fixed and the fishing club dissolved soon after the flood. For those who lived through the disaster, the dead were buried and life went on. Yet none who lived through it ever got over the flood. In the twenty-first century, even when there is no real danger of flooding in Johnstown, many still walk down to the old railroad bridge and check the river each time it rains. They have been assured that a flood like the one in 1889 can never happen again, but many still don't feel completely safe when they see the water rising.

# 10

## The Nightmare on the *Sultana* 1865

In the Nashville train wreck in 1918, more than one hundred people died because of an action that in today's terms would probably be considered manslaughter at the very least. It is certain that either carelessness or impatience caused the accident. But in the case of the explosion aboard the *Sultana,* a Mississippi River steamboat, almost two thousand people were killed because of an unthinkable combination of greed, corruption and unforgivable arrogance. It was one of the saddest tragedies in American history, and yet, by the week after the ugly events had unfolded, it was forgotten—doomed to be little more than a footnote buried in the back pages of a few books. This is one of the more horrific and dismissed calamities, as well as one of the most startling criminal acts, of the Civil War.

Lee's surrender to Grant at the Appomattox Court House on April 9, 1865, set in motion the sinking of the *Sultana.* With a stroke of a quill pen, the bloody war between the Union and Confederacy abruptly ended. Suddenly, men who had spent four years of their lives fighting battles all over the eastern half of the United States were able to look across the lines without fear. It was over and time to go home. Their prayers had been answered; they were not only alive, but they were safe.

In the Civil War's final year, the battlefields had largely been in the deep south. Death, destruction and panic had gripped the very heart of the Confederacy during this period. Hundreds of thousands of bluecoats had swarmed through cotton fields, swelling marshes, innocent towns and vanquished cities, as well as up and down rivers and along the ocean beaches finishing off an enemy that was tired, sick and often very hungry. When peace finally came the victors found themselves far from home, and getting home after such a destructive war was not a simple matter. Railroad lines had been destroyed, most horse-drawn coaches were reserved for those who had either money or position, and the annual spring floods, which always made many rivers dangerous and unpredictable, had wreaked even more havoc this year.

The most common method of transportation to the north was via the mighty Mississippi River. Four years of war, however, had taken its toll on many levees and dikes. Thus, water was over the banks for miles in the lower areas. The river was not just deep, but wide too. Trees that had been a mile or more from the riverbank were now underwater, their branches reaching up to snag anything that came by. Logs and other debris floated just below the surface, threatening to tear holes in boats that plied the river. Wrecked ships from the recent war were abandoned in the river as well. They too could shred a passing steamship as easily as a knife passed through butter. Consequently, the Mississippi had a much different personality than it had just a few years before. While it still looked calm and inviting, it was now treacherous and foreboding. Yet few noted these dangerous conditions. They were too caught up in the joy and potential that peace offered to both the soldiers who fought the battles and the businessmen whose empires were made on the river.

In April of 1865, thousands of Union soldiers gathered at Vicksburg. Some of these men had not been home in more than three years. They had fought in scores of battles, had seen thousands die horrible deaths. Clad in tattered uniforms, their spirits were beaten and their health fragile. They were exhausted. When compared to troops who had just been released from the infamous Confederate prison in Andersonville, however, the battered soldiers seemed in great shape.

Upon release from confinement in Confederate prisoner of war camps, thousands of Union troops could barely muster a victory yell. In places like Andersonville, a camp where more prisoners died than lived, the men who walked out through the gates were little more than skin stretched over fragile bones.

Located in Americus, Georgia, Andersonville was not a prison in the usual sense of the word. It was really just a stockade that covered twenty-seven acres of swampy ground. More than thirty thousand Union troops lived, suffered and died behind the camp's tall wooden fences. As the war drew to a close and Southern supply lines were cut, conditions, which had always been terrible, worsened.

Andersonville offered no shelter from the sun, wind, rain or cold. It had no fresh water. Insects not only carried disease but became a main food source. Rodents and snakes provided nourishment. Cooked rats and cottonmouths were much more appealing than the bits of food distributed in prison kitchens. Those who were held more than a few months in these conditions usually lost at least half their body weight. The inhuman conditions led to inhumane actions. It was literally hell on earth as the stronger preyed upon the weak. Andersonville seemed to bring out the worst in both the men from the North and those from the South.

When the war ended and the Andersonville gates were finally opened and the prisoners set free, they were told to go to Vicksburg. Even though the men, most weighing less than a hundred pounds, were almost too weak to stand, they actually had to march to get home. In some cases the march covered more than a hundred miles of rough roads. Using broken tree branches for canes and crutches, leaning on one another for support, getting by on one or two meals of cold beans and hard bread a day, they somehow made it to Vicksburg. When they finally straggled into the river port, the horrors of war became incredibly apparent to every person who saw these men. It would take Nazi Germany and the notorious concentration camps to again create such grim images of living humans.

As these men, looking more dead than alive, paraded into town, the regular troops stood silent and shocked. The condition of the men who lived through years in prison camps brought tears to the eyes of even the most hardened officers. No one felt anything but pity for the former prisoners. Yet what awaited them was not a heroes' welcome and a nation's thanks but a nightmare that would steal away these brave men's last piece of humanity.

Like all the war-weary men who had fought their way to Vicksburg, the freed prisoners had one thing on their minds: they wanted to go home. To them the paddle-wheel boats that leisurely cruised up and down the Mississippi were the answer to their prayers. This was what they had dreamed about for months and even years. Little did they realize that the sweetest dream they had ever had—the one of getting home—would soon become the most horrible nightmare in American navigation history.

On April 23, newspapers in Vicksburg carried the sad news that President Lincoln had been shot and killed while watching a play in Washington just a week before. The story stirred even more deeply the resentments between those from the North toward those from the South. For Union soldiers, these feelings created an even more urgent desire to go home. They not only wanted to see their families again, but they couldn't wait to get out of the South, a place they associated with great pain and suffering. Yet moving thousands of men hundreds of miles was a task that was indeed daunting. The military relied on the private sector. This initiated the problems that would soon kill a majority of those who set foot on the *Sultana*.

In order to move thousands of men back to the north to receive their discharges and be mustered out of the army, the federal government paid steamship lines five dollars a head for enlisted men and ten for each officer transported from Vicksburg to Camp Chase in Ohio. The fee established for this transport created an unexpected and welcome revenue source for some ship lines that had suffered through lean economic times during the war. Thus, the competition to haul troops opened up the door to graft, corruption and kickbacks. This would all figure into an unimagined disaster that was now just days away.

J. C. Mason captained the paddle wheeler *Sultana* and was part owner of the line that used the boat. The small company, with its fleet of four river cruisers, was in bad shape. To stay sound financially, it needed to haul as many troops as possible. As he pulled into Vicksburg on the night of April 23, Mason was intent on filling his decks with as many Union soldiers as he could find. He vowed that nothing would stop him; his ship and his company were depending on Mason to produce results.

Initially things looked promising. Mason had a big boat, a good record as a captain and knew the officers in charge of shipping soldiers to Ohio. The *Sultana* had a huge problem, however. One of its boilers was bulging. The huge bubble-shaped bulge would surely rupture if not immediately fixed. A local inspector looked at the boiler and refused to let the ship leave Vicksburg until the boiler had been overhauled. For Mason, this verdict rang like a death knell for his line and his tenure as a captain.

Mason ordered his chief engineer, Nathan Wintringer, to get the boiler fixed without delay. Wintringer sent for a local repairman, who examined the boiler and declared that it would take days to fix it. A desperate Mason couldn't afford to wait any longer than twenty-four hours, so he and Wintringer visited again with repairman R. G. Taylor. Taylor stubbornly resisted the men's attempts to sway his views. In the repairman's mind, a patch covering little more than a square yard would not do. The boiler would have to be rebuilt. After an hour of bickering, though, Taylor agreed to simply patch the bulge. When he picked up his hammer to flatten it before placing a metal cover over the bulge, Wintringer stopped him. Taylor, who knew better, was informed that the bulge didn't matter. He was ordered to leave it as it was and simply cover it with the metal patch. Taylor was then assured that Mason would have the boiler fixed in St. Louis. Though Taylor doubted it, Mason and Wintringer felt certain the patch would hold until then. Under normal conditions the patch may have held, but this was not going to be a normal run.

Taylor's quick fix took less than a day. He left the boat concerned that his work did not address the core of the problem. He had no real power to demand anything more, so Taylor knew his work was done. If the inspector now passed the *Sultana,* which he did, then the blame could not be fixed on the repairman and his work. Yet even with a boat supposedly ready to make the run upriver, Captain Mason still had trouble.

The *Pauline Caroll,* another riverboat, was now docked beside the *Sultana* and others were due at any time. The *Pauline Caroll*'s captain also wanted to take the released soldiers upriver. Rumors held that he would even be willing to pay Union officers for allowing his ship to carry several hundred of the homesick soldiers to Ohio. Now that money was involved, Mason knew he would have to counter the offer to obtain the number of passengers he wanted. There now seems little doubt that he did just that, because Captain Frederic Speed of the United States Army, the man in charge of shipping former prisoners back to the north, suddenly decided that only the *Sultana* would transport former prisoners of war. The *Pauline Caroll* and any other ship that docked in Vicksburg while the *Sultana* was there would be shut out. None of the boats received even a single Union troop.

As R. G. Taylor had been making repairs on the boiler, group after group of ex-prisoners were being marched onto the ship. They came in such large numbers and so quickly that the army didn't even have time to prepare papers for all of them. Scores were bedridden and brought to the *Sultana* on cots. Most were frail and many were sick, yet they were herded like cattle onto the boat all night long. By the morning of April 24, Mason had his cargo loaded. More than twenty-five hundred people lined his decks, the hallways, the dining room and even the area around the four boilers. It was the largest passenger load in the history of American river travel. Added to this human manifest were an additional

300,000 pounds of sugar, one hundred cases of alcohol, one hundred horses and mules and the same number of hogs.

Anyone could easily see that conditions on the *Sultana* were almost as bad as those in some prison camps. There were so many men on board that the decks were sagging under the weight. There were soldiers shoulder to shoulder on the main deck, as well as the second deck and the Texas (upper) deck. Men fought for space on the roofs, in the bowels of the ship and on crates of goods. And few of these men were in good health.

Most of those who had been marched onto the *Sultana* were suffering dramatically from malnutrition. Many had diarrhea and scurvy. They carried lice and a host of other parasites. There were no doctors or medicines placed on the ship with the troops. They were provided with no bathrooms and had no access to any cabins. They were given no blankets, in spite of the fact it was April and the days were chilly and the nights cold. Most had no coats. The one cookstove on board was reserved for the civilian passengers and the crew. The ex-prisoners were forced to subsist on salt-cured raw pork and hard bread. The water they drank came from the muddy Mississippi, which also served as the only bathroom for most of the twenty-five hundred.

Even with no place to sit down, no place to sleep, and in truth, not enough power in the engines to move this much cargo, Captain Mason pushed ahead. He ignored the way his passengers were sandwiched together and dismissed the problems he had been having with one of his four boilers. In truth, had he been given the opportunity, he would have brought more men on board. This trip was strictly about dollars and cents; compassion and good judgment had been left on the shore in Vicksburg. And though they were obviously being abused by their own government and the man in charge of the *Sultana,* the POWs were so anxious to get home that the crowded decks didn't bother most of them as much as it did the regular soldiers and civilians on board.

Though it might be understandable why Mason didn't care about the condition of the soldiers, it would have seemed that those in power at the military offices in Vicksburg would have. Many officers had been on the docks when the men had been loaded. They could see the overcrowding, they could observe the *Sultana* sink lower and lower into the river, and they knew the health problems of these men. However, they too did nothing to stop the trip. No one seemed to care much about the fate of the men who had suffered considerably for their nation.

On Monday, April 24, 1865, as the *Sultana* pushed off the dock and up the river toward Memphis, a few on board began to question if the steamboat could really manage the load. After all, it was just a typical side-wheeler. Though she was only two years old, she appeared well worked. And though she was registered at 1,719 tons, she had a passenger capacity of only 376. With more than six times that many on the boat and with a boiler that had just been patched, many figured that the *Sultana* would not have the power to make it to Tennessee, especially with the river at flood stage. But as the boat pushed against the Mississippi's current, it appeared not only strong but energetic. Within an hour the *Sultana* was making nine knots, the standard cruising speed for most riverboats of the period. As they watched the big freshwater ship plow through the muddy Mississippi, men were singing, laughing and talking about home.

The *Sultana* initially made the trip upriver look easy, but it seemed that looks were indeed deceiving. Captain Mason and his pilots were having problems steering the ship. The tremendous weight of its load had caused the boat to have an uncertain center of gravity. When a large number of passengers moved from one side to the other, the *Sultana* rocked and its heavy side dipped dangerously low in the water. Mason was concerned that the freight might slide if a majority of the men moved to one side, thus capsizing the ship. He put out an order to his passengers to stay in one place—to not move aboard the ship. It did little good.

The former prisoners had to move. Packed like sardines, they simply couldn't survive a two-day trip to Memphis without sitting down, sleeping, going to the bathroom and eating. Finally, on the second day, when the *Sultana* docked in Helena, the passengers hoped that they would be able to disembark and stretch their legs, to get some distance between them and everyone else. Except for a few crew members, however, no one was allowed off. Captain Mason took on more fuel and cargo while at port, and tragedy almost ended the *Sultana*'s journey.

A local photographer learned that the largest load of passengers ever to ride on one boat was at the dock. Grabbing his equipment, he raced down to the river to record the moment. As he set up his tripod and mounted his camera, word filtered throughout the ship that a picture was about to be taken. Many of these men had never been in a photograph, so they pushed toward the dock to be a part of the historic event. As they did, that side of the big boat leaned incredibly low in the water. The Mississippi was lapping over the decks. Meanwhile, the far side of the flat-bottomed boat was extended high out of the water. If just a few more men had crowded into the picture, the *Sultana* would have gone down at Helena. If it had, there would have probably been little loss of life. When the picture was finished and Mason watched men move back to the positions they had staked out around the boat, he breathed a sigh of relief. For the moment he and his cargo were safe.

The captain, who had seemed so confident a day before, now looked unsure as the *Sultana* pulled out of Helena. Rather than watching the ship move through the waters toward Memphis from a pilot's position, he spent most of his time at the bar. There he stood and drank one shot after another. A few of the civilian passengers became concerned as he began to walk with a pronounced wobble and his speech slurred. Still the boat steamed on and the crew seemed unworried about Mason or the big boat.

The *Sultana* fought the river without incident for two days. On April 26, 1865, at 6:00 P.M., she docked at Memphis, and those on board appeared tired but happy. The conditions were terrible, but the prisoners had learned long ago to sleep standing up and to live in shoulder-to-shoulder surroundings. Besides, the clock was ticking down—in a few more days they would be seeing their loved ones again.

Unlike at Helena, the passengers were allowed to leave the boat at Memphis. Most of them used the twenty-four hours to their advantage. The bars and saloons didn't close that night or the next morning. The whiskey flowed and the kitchens cooked everything they had in their stores. Men shared war stories and spent their money as if it were their last day on earth. For most it would be just that.

A majority of the ex-prisoners began arriving back at the dock during the early evening of the twenty-seventh. They didn't relish the idea of getting back on the boat, but they did want to go home, and they certainly didn't want to be left behind and have to wait for another ride. Still a few drank and partied until the very last moment. Some, who were obviously very drunk, were even escorted by local policemen back to the *Sultana.* Finally, with its more than twenty-five hundred passengers back on board, the *Sultana* left Memphis at 11:00 P.M.. The next port of call would be Cairo, Illinois, but before that, Captain Mason's boat would push upriver a few miles to Hopefield, Arkansas, and take on coal for the furnaces.

At 1:00 A.M., with a full load of fuel, the *Sultana* pulled away from shore for the final time and headed out into a dark and foreboding night. A cold rain was falling. Men who had been drinking in dry, warm saloons a few hours before were now freezing and wet. As the steamboat navigated waters that were more than four miles wide, thousands looked out and saw nothing. It was the darkest night any of them had ever experienced.

Samuel Clemens (no relation to the author of *Tom Sawyer*), the assistant engineer who had taken over for Nathan Wintringer at the coal station, was carefully monitoring the boilers. The furnaces, as well as the boilers themselves, were red hot. The metal was glowing. A man could only stand within a few feet of the furnace without cooking his skin, and the troops who had been forced to ride in the boiler room were growing very uncomfortable. As Clemens studied his boilers, a well-known physical fact must have entered his head. If just one boiler blew at the pressure they were running at this moment, the resulting force of the explosion would carry pieces of steel more than two miles. To make matters worse, these metal fragments would no doubt puncture the other boilers and set off a chain reaction.

Clemens, who had been concerned about the repaired boiler since leaving Vicksburg, stood beside the huge metal cylinder and examined the patch. It seemed to be holding. There was no evidence of any leaking of steam or water. Yet the bulge looked bigger. Was it? Or had it just been magnified by his concerns?

Well above Clemens stood George Kayton. Another man assigned to the night shift, he was at the *Sultana*'s wheel. He was responsible for guiding the boat up the swollen river, avoiding all the hidden sandbars and dangerous eddies. He could see little in the darkness. The rain had streaked the windows all around him, so Kayton was largely going on feel. If the ship felt right, he figured he was in the right channel and safe. If the movement varied a little and the feel of the wheel seemed different, he made minor corrections until he judged the feel to be right again. It should have been a nerve-wracking job, but the pilot was experienced and sure that he could take the ship safely through the night.

As the *Sultana* made nine knots an hour, Captain Mason again excused himself to the bar. There he drank glass after glass of the ship's finest. As the booze began to rob him of his inhibitions, he began to voice his own concerns about the boiler. He told several of the civilian passengers that he was worried. "I just hope the boat can hold together until St. Louis," he was later quoted as having said several times. He also informed all who would listen that he planned to have the boilers reworked in the proper fashion when they docked in St. Louis.

In comparison to the *Sultana,* the *Titanic* was well equipped for disaster. The riverboat carried just one lifeboat. It was very small and would hold only a few people. Another larger boat rested on the decks, but it was not meant for use during disasters. It mainly served as a way to ferry goods and passengers from the steamer to the shore. The *Sultana* was only outfitted with seventy-six life preservers. There was no room for error on this boat. As the sidewheeler reached a point seven miles north of Memphis, the lack of life-saving equipment would come to mean a great deal to thousands of unfortunate people.

By two in the morning, most of those on board had gone to sleep. The troops were lying haphazardly across decks and even on roofs. They were literally using each other for pillows and squeezing closely together to fend off the cold and rain.

As Mason drank and Kayton steered, engineer Clemens again checked the repaired boiler. At that point it really did seem to be bulging more. On top of that, it appeared as though steam was escaping around the patch. The assistant engineer didn't have time to notify anyone of his observations. At precisely two in the morning, the boiler gave up.

One second the river had been silent except for the sound of the *Sultana*'s big wooden paddles thrashing through the water. The next second, a tremendous explosion that could be heard and felt in Memphis sent an orange-colored flame boiling into the black sky. At the same moment, a huge pillar of fire suddenly lit up the black, swirling river. It looked as if a volcano had erupted in the middle of the mighty Mississippi.

The leaky boiler probably blew first, but a millisecond later the boilers on each side blew as well, probably caused by debris piercing them. It was a miracle the fourth one somehow avoided ignition. With water, steam and hot pieces of molten iron leaping skyward, another of the *Sultana*'s weaknesses came to light.

The boat's decks had been constructed of the very lightest wood. The decks had little bracing. The force of the exploding boilers literally blew the decks into tiny splinters. What it did to the men who were sleeping on those decks above the boilers was unthinkable. Many were stabbed with thousands of little pieces of wood, others had their bodies blown in several different directions at once. The lucky ones were simply tossed into the air, many coming down into the river more than a mile from where the *Sultana* now burned.

Hundreds of those who managed to avoid becoming human cannonballs were simply cooked. Boiling water and steam came up from the bowels of the ship like lava. It shot into the air, then splashed down on men who a second before had been sleeping. They had no time to react, no time to move, no time to pray.

Fire was everywhere, and less than half a minute after the initial blast, the whole ship began to slide down into the lower decks. Horses, pigs, mules, cargo and even men were being sucked into the fiery furnaces that were now burning out of control.

Hundreds who had not been instantly killed or swept into the fire were floundering in the water. Still more asleep than awake, they fought the river like they once had the Southern army. It was hand-to-hand combat, but the muddy Mississippi had the clear advantage. Not only were the men in shock, trying to survive in the dark, but few of them knew how to swim. In 1865, only one person in ten in the United States could even tread water.

Weakened by the time spent in prison camps, hundreds were now shoulder to shoulder in the muddy water beating each other to death. As the men fought the river, they also fought anyone around them; clinging to anyone who seemed to know how to stay afloat, survivors were pulling others to watery graves.

Those still on the burning ship looking out across the Mississippi saw masses of hopeless humanity cursing, crying, praying and begging. They were also grabbing anything they could find. As panicked witnesses watched the life-and-death struggle in horror, one fact quickly became apparent: The strong were destined to survive and the weak were doomed to die. Few in the water had much strength and it seemed very few would live more than the time it took to wave good-bye.

Within minutes of the explosion fewer than five hundred people were left on board. As they looked at the doomed ship and the equally frightening water, many knew they were going to die. Some even took a few moments to pull out paper and lead and scribble notes to loved ones—the same loved ones they thought the *Sultana* was returning them to. They carefully signed their final letters and placed them in their pockets. For many, these notes would serve as the only way their bodies could be identified.

On the *Sultana,* scalded men pleaded for mercy. Those who couldn't get up from cots begged to be tossed to their deaths in the cool water. And the heat from the fire grew worse and worse. As the flames spread, those left on the ship realized that they too would have to join the teeming masses in the river. Grabbing doors, bales of hay, the bodies of horses and anything else they could find, they jumped into a cold Mississippi that was filled with scores of living men, as well as severed limbs, heads and torsos. The water seemed much more inviting than did a boat that was now consumed by the smell of burning flesh.

The *Sultana,* now a flaming mass of wood, floated helplessly downstream back toward Memphis. It seemed to be chasing the very men and women it had just ejected. As the ship approached them, its flames lighting their faces, hundreds of people screamed out for help, but help for most was not to come.

Those who managed to swim to the Arkansas shore climbed up in trees. As they did, mosquitoes swarmed all over them. Hypothermia set in as well. Many who had survived the blast and the swim to safety died in the trees and slid back into the water.

The few people who witnessed the blast sent word of the tragedy via horseback to Memphis. Union ships ordered out their small boats, and other steamships, docked at the port that night, also went out into the water. Many farmers lashed logs together and took to the river in homemade rafts. Yet with no lights, on this rainy dark night all of the would-be rescuers had to carefully listen for cries or react quickly when their boats hit a body. In a testament to their fortitude and courage, some did manage to save a few victims.

The *Bostonia II,* another riverboat steaming in from Cairo, sent out a boat, and its crew managed to pull more than twenty barely living men from the river. Dewitt Spikes, who had been blown off the *Sultana* and had swum to safety, secured a boat and went back on the river. He pulled in more than thirty people that night, but could not find his father, mother and three siblings. They drowned, leaving the teenage boy an orphan.

Two Confederate veterans pulled more than a hundred Union

troops out of the water and took them into their Arkansas homes. There they were fed, given medical care and placed beside warm fires.

Most who went out to help those struggling in the river were too late. Injuries, the cold and the water got the victims before rescuers could.

As the *Sultana* burned and drifted, almost thirty men remained on board. Most were too sick or too scared of the water to jump from the wreck. By the time the steamship finally floundered in an eddy just off Chicken Island, the flames had pushed these men to the last part of the forward deck. They would have faced certain death if riverman John Fogelman, using a homemade river raft, hadn't fought the Mississippi currents and brought them all to safety.

When the cold dawn light broke through the dark night skies, survivors dotted the river all the way to Memphis, clinging to logs, rafts, barrels, sections of railing and even dead horses and hogs. By mid-morning, 750 had been plucked from the waters. More than 500 men were taken to the Memphis hospitals. Some 200 of those joined 1,700 others in the arms of death. Of the 23 women on board, 22 died. Somehow 783 survived: 765 troops, 18 civilians and crew members.

The sunlight revealed a grim picture. The dead littered the shore on both sides of the river, caught in trees and brush, some floating intertwined as if a large human raft. Bodies were found along the river for weeks. Crows and hogs, as well as dogs and other wild animals, fed off the bloated and beached corpses. Some of the dead were caught in steamboat paddle wheels. Most of these victims were simply yanked out like driftwood and pitched back into the river. The dead floated to Helena, back to Vicksburg and even beyond. After a few days, spotting a body in the Mississippi south of Memphis was such a normal occurrence that few treated the event with any reverence or the body with any dignity. One of those who suffered this fate was Captain Mason. His body, like hundreds of others, was never found.

At first it was thought that sabotage had caused the calamity. That is what most had hoped happened. It would be easy to blame things on the South and the few Rebels who would not give up. Yet it quickly became apparent that no enemy had hit the ship and the Confederacy had no stomach for killing helpless former prisoners. It was then that the official inquiries questioned the condition of the boilers.

Three of the boilers had been blown to pieces in the blast. As they were of a new design, there was very little history to study to help determine what had gone wrong. A commission decided that the boilers had probably been overworked, had not been properly cleaned, and that at least one needed to be replaced. But it refused to cite the steamship company for their failure to address these problems. The survivors and families needed someone to blame for that "accident," the commission still needed to find a person to whom they could point. The logical choices would have been the *Sultana*'s engineer and the captain.

The army, however, also quickly cleared Nathan Wintringer, who was one of the fortunate men to survive the explosion. For some unknown reason, they didn't make Mason a scapegoat, even though he had died. Those in charge decided to hang cause for the blast on Samuel Clemens. Even though Clemens probably didn't have the power to do so, the examining board determined he should have asked for the ship to stop and shut down the faulty boiler. Besides, as he was dead, he could not protest the finding.

The United States Military commission also absolved the army of any wrongdoing. Even though several officers testified that Fredrick Speed had been bribed, the commission decided he had been guilty only of poor judgment. Speed would go on to represent Mississippi in Congress.

When all the investigating was finished, only one conclusion should have resulted, but it was not a part of most official records. The Band-Aid repair on the boiler, which used very thin metal, was simply, as the repairman had argued during the repairs, not up to the job. It would not stand the pressure needed for riverboat travel. Experts felt that the patch could hold no more pressure than just over 100 pounds per inch. The *Sultana* was maintaining more than 150 pounds per inch on its final journey.

Newspapers outside of the Memphis area all but ignored the story even as it happened. A handful devoted a few paragraphs to the disaster. Why was it dismissed so easily? The calamity had happened in the west (now the midwest) and most newspapers in America were in the east. At that time newspapermen didn't view stories from the "frontier" with much interest.

It was also worth noting that the casuality figure was not that shocking to Americans in 1865. While almost 2,000 troops had died when the *Sultana* exploded, this was a drop in the bucket compared to the 600,000 who had perished in the war itself.

The most important reason the story was not a worldwide news event is centered on a tragedy that overshadows it to this day. The assassination of Lincoln and the search for and then the killing of John Wilkes Booth was a story that generated far more newsstand attention and spoke much more to the heart of the American public.

Even in Memphis, the *Sultana* was quickly forgotten by the newspapers. As death wagons hauled a dozen bodies at a time to cemeteries, no one seemed to notice or care. Even when victims' relatives placed heart-wrenching ads in the papers as they searched for bodies, few read them. This may be the reason that many pulled from the river were never identified.

The *Sultana* should have been recognized as another tragedy signifying the futility and horror of the Civil War. This wreck should have changed the way riverboat passengers were treated and the way business was conducted on the nation's waterways. For the sake of safety, changes should have been ordered and made, and the men who needlessly gave their lives in order to pave the way for safer travel should have been lionized. Yet, as it was, the events, greed and irresponsible actions that led to this disaster were whitewashed. And the victims who survived not only war but inhumane treatment in prison camps—the very ones who thought when they boarded the *Sultana* their prayers had been answered and that they were finally safe—should have lost their lives for something far more noble and meaningful.

# 11

# The Port Chicago Nightmare 1944

The African-American men stationed at Port Chicago, in San Francisco Bay, had not enlisted in the navy to simply work on docks in the United States. They wanted to fight for their country. They wanted to meet the enemy head-on. Many of them had been inspired to join the service by Doris Miller's heroics at Pearl Harbor. These men wanted to do for America and their race what Miller had done.

On December 7, 1941, as scores of Japanese planes wrecked the Hawaiian base and killed more than two thousand, Miller was on the *West Virginia* working as a cook. Because he was black, the Waco, Texas, native was not expected to ever see battle or fire a gun. Even in the armed services, Miller was considered a second-class citizen.

While Miller worked in the galley that fateful day, enemy fighter planes took out the crew manning one of the ship's antiaircraft guns. With no instruction and no help, Miller pulled the dead and injured men out of harm's way, then stopped and studied the weapon. Using his instincts and common sense, he figured out how to load, aim and fire. Before his ship was blown out of the water, Doris Miller would bring down four Japanese Zeroes. One of the few Americans to score a fatal hit on the enemy at Pearl Harbor, "Dorie" was rewarded with the Navy Cross.

Miller was brought back to the states to recruit other African-Americans into the service. For blacks in America, having one of their own be recognized as a hero was a new occurrence. Scores of young black men began to see the navy as a place where they could be judged on merit and not by the color of their skin. They signed up hoping not only to fight enemies in Europe and the Pacific, but to fight racism in the United States. Surely, if enough black men distinguished themselves in battle like Doris, then society would be forced to open its doors of opportunity to men of color when the war ended. What these men didn't know was that when Miller returned to duty, the sharpshooter did not man a gun; he watched the war from behind the cook stove. Many other blacks would find themselves limited to similar positions. As it turned out, the ones who worked in the kitchens, even in battle zones, were safer than those who drew duty on the Port Chicago pier.

The men who found themselves assigned to the Bay Area docks could not believe they had endured weeks of basic training in Great Lakes, Illinois, to spend their days on a small base in Port Chicago, California. This tiny town, with a few stores, a movie theater and a couple of cafes, offered little excitement for any man in uniform. And for the hundreds of blacks who were stationed here, the town offered nothing at all. Port Chicago was a "white only" community. African-Americans could not shop in the stores, watch a movie or buy a meal. For many of the young enlisted men, the base was not only thousands of miles from the war, but a prison without bars. Just as it had been when they were civilians, race again held them back.

Not only did the black servicemen not want to be in Port Chicago loading munitions, but their base commander, Captain Nelson Goss, did not want them there either. A graduate of the United States Naval Academy, Goss viewed blacks as untrustworthy, shiftless, immoral, and lazy. He stated his convictions in both conversations and letters to superiors. He felt that "negroes" did not possess the intelligence to even load ships, much less participate in the other, more important duties of a navy seaman. Goss requested several times that the African-Americans at Port Chicago be reassigned and replaced with white sailors.

The local civilian dock unions had similar feelings, but unlike Goss, their complaints about the men stationed at Port Chicago had nothing to do with race. The experienced longshoremen felt that no man, white or black, should be loading munitions off of railroad boxcars and onto ships without extensive training. The union noted that their crane operators had to work years, display great talent and have a perfect safety record to work with explosive materials. These veteran longshoremen were horrified that the navy would build a facility such as Port Chicago so close to a civilian population and then staff the facility with untrained workers. The union men were afraid that a calamity of horrendous proportions was just waiting to happen. Yet their warnings went unheeded.

The coast guard also voiced concerns. Like the longshoremen, they knew of the lack of training and felt that the operation would be more successful if their own men oversaw the work at the docks. The navy did not follow the coast guard's suggestions either.

There can be little doubt that the war played a part in the navy's lack of concern over the safety issue. To worry about men loading ammunition seemed rather silly when thousands were dying each day on the battle fronts in Europe and Asia. Most officers saw any job, even the ones at Port Chicago, as a walk in the park compared to being in combat and few officers really understood the dangers of loading munitions in ships. They had as little training and experience for this assignment as the men who worked under them.

The black men assigned to Port Chicago were naturally fearful and apprehensive when they learned of their assignment. Most had been trained to staff kitchens, not unload explosives, and a few still dreamed of an opportunity to prove themselves like Doris Miller had, if only by accident. But outside of San Francisco, thousands of miles from a war front, they found themselves loading the bombs, bullets and arms that would kill the enemy, not delivering those death blows in person.

The irony of being entrusted with a very dangerous job so far from the war was not lost on the men. They spoke of it often, knew that they were doing this job not because of training or expertise but because they were considered expendable. If something happened, the navy figured they would have no trouble replacing them. It was disheartening to realize that they were not considered worthy of doing the same jobs as the white men. And morale was low.

For two years, since the day Port Chicago opened, black men in white uniforms had loaded bombs and other explosives twenty-four hours a day seven days a week. Throughout this entire time, they were given no formal instructions on how to lift, stack or sort materials. They simply learned as they went along. With white officers acting like prison guards watching their every move, Captain Goss and his chalkboard keeping track of each day's work, the men were continually pushed to load more. By 1944, when the United States and its allies finally were gaining the upper hand, working at the pier was not so much a job as it was a race. Unlike the war, this was a race with no end in sight.

As the daily grind of the work took hold, the men in Port Chicago realized that they were in a grim situation. They were ridiculed because they could not work fast enough, were called lazy and shiftless, and were constantly berated because of their race. Even when they set records for production, they were told that they did not work as well as white men. Furthermore, they faced a future with no promise of advancement through the ranks.

White officers, during meetings with the black seaman, assured the men that they were safe. The bombs, many weighing more than two thousand pounds, could not explode because they had no fuses. Neither could the artillery shells. Even though they were loaded with TNT, they came in on the trains as duds. They had no detonaters or fuses. Therefore, no one had to be especially careful, just quick.

Divided into different groups and shifts, the squads who produced best were rewarded with passes to San Francisco or Oakland or extra time for writing letters and doing laundry. Over time competition between groups anxious to get away from the docks raised the amount of munitions being loaded and lit the fuse for an explosion that had to happen.

Longshoremen, who came by from time to time to observe conditions at Port Chicago, shuddered at the way the munitions were being handled. Inexperienced crane operators were banging bombs against boxcar walls and the men loading the ships were tossing the explosives around like super-heavy basketballs. Metal was clanging against metal in the ships' holds, on the docks and the trains. Yet as time went by and none of these actions created any problems, the seaman began to actually believe that the work they were doing was indeed routine, although the long hot hours of backbreaking labor still took its physical and emotional toll.

Many of those at Port Chicago knew that they could be working on a civilian industrial job for many times the money they were making in the service. For volunteers this was especially disheartening.

Requests for transfers to the fronts were common. But the navy wanted the black men right where they were.

On the evening of July 17, 1944, there were more than fourteen hundred black enlisted men divided into eight different divisions at Port Chicago. Each of the divisions worked together on the same shift and the same jobs. In charge of each group was a white officer. Along with the African-American workers, there were more than 70 officers (a few of them black petty officers), 106 marines and an additional 200 civilians who worked at such skilled positions as carpenters, locomotive engineers and repairmen.

Many of the black workers who had finished their shifts were resting in the barracks on a segregated part of the base on that night. In this "colored only" section the seamen were thinking about the war in Europe. D day was still fresh in their minds. When they had signed up, many thought they would be a part of it. Now, as the cool salt air drifted lazily through their rooms, they realized that they would never be allowed to fight for their country. This caused many to wonder if they were really a part of the nation where they had been born and raised.

Those in the barracks were lucky, their hard shift in the holds of the ships and on the pier was over. It was much more difficult to work at night when the lighting in many areas was not that good, so there was less room for error. The night crews were expected to produce more than those who worked under the hot sun, so most dreaded pulling night duty.

On the evening of July 17, it was very dark. There was no moon at all. The breeze had brought in cool air from the bay, however, so except for the blackness of the sky, it was not a bad night to work.

There were two ships ready for loading at the Port Chicago pier that evening. The Liberty ship SS *E.A. Bryan* had been on the landward side of the pier for four days. Around the clock men had taken munitions off of freight cars and placed them into the holds of the huge cargo ship. By ten, the *Bryan* held over forty-six hundred tons of ammunition and explosives. It was almost full. Ninety-eight untrained black enlisted men, a part of the Third Division, were working at a reckless pace trying to finish loading the ship before morning.

Thirty-one merchant marines watched from the ship. They were as anxious to get out of the port as the workers were ready for them to leave. The seamen felt it was much more dangerous being tied to land than sailing on the open ocean. Here they had little control over their ship and were dependent upon the navy to do the job right. Many of these men had spent years watching experienced longshoremen load volatile cargo and realized that the hardworking black sailors had their hands full with no real training. The merchant marines were betting that disaster would strike the port someday, they only prayed it didn't happen when they were tied up at Port Chicago.

Showing much less concern than the seamen were the thirteen naval armed guards. They were on the pier to guard against acts of sabotage. As there had never been a single act of stateside sabotage in the entire first three years of the war, these "lucky" thirteen were not too concerned about their safety on this night or any other as long as they were stationed at the pier.

On the other side of the *Bryan* was a brand-new ship. The SS *Quinault Victory* was on its maiden voyage. It had only tied up at the dock four hours before. The Sixth Division had been assigned to christen the new vessel with its first cargo. More than one hundred men assigned to the job were inspecting the ship at ten before beginning to load her around midnight. Thirty-six merchant marines were showing the black sailors around the *Victory.* Casually observing the process were seventeen more navy guards.

Also at Port Chicago that night was a coast guard fire barge. The few crew members on board were getting ready for bed. Some visited with the sailors, many of whom they had gotten to know well, but most of the small body of men simply observed with disinterest the activities of what appeared to be another normal duty shift.

On the docks, over 430 tons of bombs sat, silently, waiting to be loaded. Scores of men scrambled all around and over these weapons of war as they went about their duties, each of the sailors seemingly oblivious to the power locked in each of the metallic cylinders. The bombs had been stacked on the pier in order to speed their loading. When the signal to fill the *Victory* was given, these would be the first things loaded.

Beside those bombs were sixteen boxcars waiting to donate their volatile cargo to the war effort. Sitting in front of the cars was a steam locomotive. The engineer was looking forward to getting out of the port and away from the dock. Like the men who loaded the ships, the explosive potential that was all around him made the man nervous.

Work went well that summer night, and on breaks the African-Americans could take some pride in how quickly the *Bryan* had been loaded. As they rested, they spoke of baseball, sweethearts, family and news from the battle front. A few played cards, others drank coffee and cokes and read letters from home. The one thing they didn't do was smoke. It was about the only taboo for those working on the dock that night.

Before the break ended and the crews went back to work that night, a white officer visited with a concerned sailor who was new to the port. The young man, fresh out of basic training, was scared that the bombs he was helping to load might explode. After explaining why it couldn't happen, the officer added a grim postscript: "Besides, if something did happen and a bomb did go off, you'd never know about it! You wouldn't feel a thing!"

The spark that brought these words to life began on the *Bryan* just after the men returned to work. The ship had five separate holds. In the last one, number five, forty-millimeter shells were being loaded. Cluster bombs were filling up number four. The next was home to thousand-pound bombs. The number-two hold contained depth charges. The first cargo hold was being loaded with incendiary bombs. This is where "the fuse was lit."

Cargo hold number one was filled with "hot cargo." The six-hundred-pound incendiary bombs were complete and ready to turn loose on the Japanese. Unlike all the other munitions being loaded, the incendiary bombs' fuses were in place. If one of these was dropped too hard against another one, there was a chance of an explosion. Each time metal clanged on metal in this hold, sailors paused and prayed. So far their prayers had always been answered.

The crew working in number one was moving much more slowly than the other crews. They were carefully taking each incendiary bomb from the crane's net and slowly "dropping" it into place alongside hundreds of other hot bombs. Even in the cool night, these men were sweating, and unlike the other crews, they rarely talked as they worked. They treated every moment as if it was their last.

Lieutenant Commander Glen Ringquist was in charge of the operations that evening. Satisfied that all was going well, he left the pier just after ten to file his report at the administration building. He knew that Captain Goss would want to know if the men had met their production quotas for the evening. Ringquist also knew that if the quotas were met, then they would be raised and the men would have to work faster and harder during their next shift. There would be no pleasing Goss, so the sailors who drew duty here were doomed to working like slaves for a taskmaster who would never believe in their value or reward their efforts.

At 10:17, everything was normal. The loading of the *Bryan* was almost complete and the process of filling the *Victory* was about to begin. In the background, just to the side of the pier, a crew was hurriedly unloading boxcars, while on the dock itself men were reaching for the incendiary bombs that would finally top off cargo hold number one. It was a hauntingly peaceful scene, one that showed an orderly and efficient navy crew working on an eerily quiet night.

At 10:18, those in the barracks more than a mile away from the pier had switched off their lights and were drifting off to sleep. The few men in the administration building were silently filling out their reports and answering occasional phone calls. High above them an army air force crew, flying at nine thousand feet, looked down at the harbor as their plane buzzed through the dark sky. On the dock itself, men were so caught up in their normal duties that few even noted the drone of the passing plane.

Then the unthinkable happened. In an instant, what had been a peaceful summer evening erupted into a massive ball of flame that seemingly shot straight from hell. It happened so fast that it was unlikely anyone on the Port Chicago pier even realized what hit them. James Born, Leroy Hughes, Lawrence Jackson and Isaac Smith never had the chance to look up or take a last breath. Willie Nettles, Gabe Johnson and Cluster Hill might have heard a metal clang, but nothing more. In a second these seven and scores of others were gone.

Just after those around the dock heard the metal ringing, the army air force crew heard a rumble and felt the air grow suddenly turbulent. As they held tightly onto their controls, they were shocked to see pieces of white-hot metal, some as large as houses, zoom past them. According to the copilot, the "fireworks display" lasted about one minute. It was a bizarre sight, as if the bowels of the earth had suddenly opened, intent on burning everything in sight. Somehow the plane dodged the debris and flew safely across the bay.

Those on duty at the University of California at Berkeley watched their seismographs move twice. The men thought they had just witnessed an earthquake, not large, but possibly locally severe. It would be hours before they discovered the origin of the "quake."

At that same moment, windows blew out in the barracks and administration building, walls and roofs collapsed and men were tossed around like rag dolls. Many were killed instantly, others were so badly injured that they didn't have time to consider what might have just happened. A few who escaped any real injuries believed that the Japanese were repeating the attack on Pearl Harbor, this time with San Francisco as the target.

Crawling out from the rubble, those who had not been blinded by flying glass were greeted by an unimaginable sight. A column of fire more than a mile high was shooting straight up in the air at the dock. Smoke, billowing ahead of the flames, seemed to move at supersonic speed. The cool night air was suddenly hot. It was then they realized that their worst nightmares had come true.

The explosion tore apart the base as if it had been constructed of cardboard and rocked the bay like a tidal wave. A wave of more than forty feet hit the *Miahelo,* a coast guard patrol boat, tied up about fifteen hundred feet from the Port Chicago pier, and nearly capsized the ship. Many of its crew were badly wounded by flying metal, some the size of cannonballs.

Not far away, a sixteen-inch shell that somehow did not explode hit the engine room of the small tanker SS *Redline,* anchored several thousand feet away from the pier. The seasoned sailors thought they were under attack as munitions, some exploding, others just thousand-pound duds, were flying in all directions.

In the city of Port Chicago, hell was raining down as the people were being hit by their own navy. Bombs and pieces of ships were raining down on the tiny community like fiery hail stones. Buildings were collapsing, windows were being shattered, fires were popping up on roofs, in yards and in streets. Cars that had been parked in front of homes were now nowhere to be seen. Nearly all the town's thirty businesses and public buildings were destroyed, and no home escaped without some damage. As the sleepy west coast town awoke in horror, the cool night had suddenly grown very hot. As badly as the explosion wrecked the town and base, it was at Port Chicago where the greatest damage was done.

The twelve-hundred-foot-long wooden pier, the locomotive and boxcars, the SS *E.A. Bryan,* and 320 people, 202 of them black enlisted men, were instantly gone—literally vaporized. All sixty-seven crew and thirty armed guards aboard the two ships died too. Only fifty bodies would survive in a complete enough form to identify.

It would take almost half an hour to get ambulances to the scene. Initially it was too hot to get to the heart of the explosion, but there were hundreds of men, some who had been more than a mile and a half from the explosion, who needed medical attention. Exhausting the supply of ambulances, buses were called in to transport the wounded to hospitals.

Black sailors, most with no firefighting experience, grabbed hoses and rushed as close as they could to the flames. As they marched behind the stream of water, they were horrified to see that several boxcars of bombs were on fire but had not yet exploded. Rather than turning and running for cover, five of these men rushed to the cars and doused the flames. This heroic act probably saved several hundred lives. It was the first time any of these black enlistees had been given an opportunity to have their "Doris Miller" moment. In the face of death, they proved themselves as Doris had in a time of decision and action.

When dawn's light touched the bay, no identifiable pieces of the SS *E.A. Bryan* remained. The stern of the SS *Quinault Victory* lay upside down in the water five hundred feet from its origin. The rest of the ship was in scattered pieces. In little more than a second, 25,000,000 pounds of ship and ammunition disappeared. The force of the explosion was on the same order of the bomb that was later dropped by the Americans on Hiroshima.

As the unhurt seamen began to clean up the rubble that had once been a naval harbor, they found not only huge pieces of unexploded bombs, but body parts, many too small to even identify as being human. As they picked up heads, hands and feet of men they once knew well, the horrors of the event became too much for many of them. White and black sailors alike cried like babies as they considered what these men's last second must have been like.

Days after the explosion at Port Chicago, James Camper, William Anderson, Rickel McTerre, Effus Allen and John Haskins were awarded the Bronze Star. Rear Admiral Carleton H. Wright stated, "I am gratified to learn that, as was to be expected, Negro personnel attached to the Naval Magazine Port Chicago performed bravely and efficiently in the emergency at that station last Monday night. These men, in the months that they served at that command, did excellent work in an important segment of the District's overseas combat supply system. As real Navy men, they simply carried on in the crisis attendant on the explosion in accordance with our Service's highest tradition."

With the actions of these five, and the bravery of those who acted without training on that night, the African-American sailors' fight to win equal footing with their white comrades should have been secured. Who could doubt these black men's courage and abilities as soldiers now? The answer would be evident just as soon as the navy commission assigned to investigate the disaster made its report.

The judge advocate of the District Intelligence Office wrote, "The consensus of opinion of the witnesses—and practically admitted by the interested parties—is that the colored enlisted personnel are neither temperamentally or intellectually capable of handling high explosives." In other words, the navy hung the disaster and the deaths on the untrained black men who had been killed in the explosion. The commission echoed the thoughts of Captain Goss.

Ironically, though judged to be the reason for the disaster, hung with the blame and cited as being intellectually incapable of handling explosives, the men were given the same duty again, this time at Mare Island. The officers making the assignments didn't believe the finding of the commission or they didn't care if more black men were being placed in a position where hundreds might die the same way those at Port Chicago had.

The remaining black ammunition handlers, many of whom had been quietly voicing concerns about safety for some time, naturally feared loading ammunition again. When more than two hundred refused to go back to work, the navy acted. The same service branch who had just stated African-Americans could not perform as munitions handlers decided to punish those who refused to load the boats. Fifty enlisted black men were singled out and tried for mutiny, including one with a broken arm. All were found guilty and sentenced to fifteen years of hard time. The navy court never saw the irony that the verdict created. Its own officers had declared the men incapable of doing a job, then, when the men agreed and refused to do it, navy officers declared them to be cowards and enemies of the service. The truth was that no black or white man without special training should have been involved in loading munitions. As civilian longshoremen had pointed out time and time again, it was simply not safe.

In January 1946, after spending more than a year in confinement, the "mutineers" were released from prison but forced to remain in the navy. They were sent to the South Pacific in small groups for a probationary period, then gradually released with dishonorable discharges. Being branded as traitors would follow many of these men for the rest of their lives, as would the nightmare of the disaster itself.

For those who somehow survived the blast at Port Chicago, the same men who thought they had been given a chance to fight for their country, World War II only served to remind them that they were considered second-class citizens. As millions marched home from war feeling safe for the first time since 1941, these men grew to fear not the enemy across the seas, but the ones who were in charge of the military services that were supposed to ensure their freedoms. Only now, almost sixty years later, are the few survivors of the tragedy at Port Chicago being recognized for their courage in not just working at the munitions dock, but in standing up to a government's injustice.

# 12

# WACO

## The Texas Twister

## 1953

On May 11, 1953, the eighty-four thousand people who called Waco, Texas, home woke up to a beautiful day and some very tragic news. Headlines in the *Waco News-Tribune* announced in large bold print that six members of a local family had died in a tornado near Minneapolis, Minnesota.

With great interest thousands in the small, tight-knit Texas community hurriedly scanned the paper for the particulars of the tragic event. They discovered that eight members of the Martinez family had recently traveled to the Midwest to work as migrant vegetable harvesters. They had been living in a small frame house when the twister appeared without warning, literally blowing them out into the very fields where they had labored only the day before. The tragedy left the injured mother to contemplate how she would ever be able to go on without all but one of her children. Storms, it seemed, had no respect for property or family. They also had little regard for location or legends. Waco would soon learn these important facts in a very devastating manner.

As the news of the Martinez tragedy filtered across the city, collections were set up to aid the family with burial expenses, transportation home and medical costs. Yet, even as many locals reached into their pockets and purses to help their former neighbors, few believed that they would ever face a similar catastrophic event.

As the unusually warm and humid day passed, as morning gave way to afternoon, those in coffee shops and around office water coolers may have understood the fragility of life as witnessed by the Martinez family, but they didn't relate those events to any in their own lives. After all, not only had the weather bureau promised them no storms that day, but those in Waco knew they lived in a city where tornadoes never hit. They had every reason to believe they were truly safe. History was on their side as well.

A thousand years before Columbus decided to prove the world was round and stumbled into America, the Huacos Indians had moved to the area along the Brazos River now known as Waco. These Native Americans had flocked to this Central Texas spot and established a village along the stream not because of its beauty, rich soil or plentiful game, but because they believed that deadly twisters could not touch this place. They thought that the gods had placed this area off limits to that kind of destruction.

This faith never wavered, nor did the Indians' love for Central Texas. Yet in 1953, they were gone. The Huacos were not routed out of their homes by a tornado—the cyclone that cost them their way of life was a group of white men who had decided that they needed the storm-free land much worse than the Indians. In their minds every Native American belief was childlike and ignorant, other than the one that logic should have told them was the most outlandish. In short order the settlers nearly wiped out all that had anything to do with Waco's original citizens, except the legend that the ground around the city was immune to tornadoes. Even in the middle of the twentieth century, a majority of Wacoans still hung on to the belief that their hometown was immune to tornadoes. Some weather experts, ignoring all training to the contrary, explained that the city was protected from twisters by the hills on the west, south and north of town.

Even though a tornado seemed out of the question on this Monday, the local weather service had predicted a chance of rain late in the afternoon. So a number of folks were trying to get their business taken care of before any of the stronger showers hit. At four in the afternoon, when the lingering showers began to gather some force, downtown Waco was still filled with thousands of people apparently enjoying a rather soggy May day. By this time the rain had become a nuisance and had created a rush on umbrellas at Montgomery Wards, but it didn't appear menacing.

Like much of America, even when it was raining Waco was filled with post–World War II optimism. Once almost completely dependent upon agriculture, the community had suffered greatly during the decade before the war. The Depression had almost brought the area to its knees and taken the city with it. But now, as Americans geared up to lead the industrialized world, Waco had an energy and hope that fueled growth and good times. The Central Texas area was again a happy place to live and work. And Waco, as much as any community in America, combined a family atmosphere and old-fashioned values with a new spirit and can-do attitude.

As in most cities of the time, the downtown area of Waco was where everyone came, especially on Mondays and Fridays. It was where people ate, played, conducted business, visited government offices, attended schools and went to the movies. When people spoke of Waco, they generally thought of the picturesque downtown area.

In the middle of long rows of beautiful brick structures stood the five-story Dennis Building. It housed a furniture store, but it was much more than a center of commerce, it was a place where people caught up on all the local news. Unlike many southern businesses of the era, this family-owned establishment catered to every race, even employing Hispanics and African-Americans in a wide variety of positions. During this May afternoon the Dennis Building was filled not only with those looking for a new mattress or chair, but with refugees from the rain. The store's huge showrooms and plate-glass windows were perfect places to stay dry and watch the rain pepper down on Fifth Street. As almost a hundred waited out another downpour, conversation centered around the fate of the poor Martinez family.

Behind the Dennis Building, just across the small Banker's Alley, was a recreation hall. When school let out, this combination pool hall and domino parlor was one of the most popular places in Waco. Across the street from the Dennis Building was the Joy Theater. As was the case with the furniture store, a host of folks had decided to escape the rain by ducking inside to catch a show.

Down the next few blocks were other multistory structures, beautifully designed brick edifices complete with awnings and huge plate-glass windows. Some, such as the Professional Building, were largely filled with medical offices. Others, like the Padgitt Building, once a saddle store, were now home for shops and professional offices. On the other side of these two landmark buildings were several blocks of active commerce that included the Gem Theater, Sammy's Cafe and the great Alico Building, a twenty-two-story marvel that had been the first skyscraper in the western part of the United States. From the top of the Alico, a person could see everything in Waco, from Baylor University, to the farmer's market, to Katy Park to the Waco Suspension Bridge. Yet from that perch on May 11, 1953, in the late afternoon, an observer could have seen something else too. Something that was not supposed to be there.

In spite of the legend, a few local citizens had not liked the look of the sky that afternoon. They knew from news reports that a tornado had struck San Angelo, a couple of hundred miles to the west, earlier in the day. They were wondering if perhaps that storm was coming toward Waco. Yes, they were told by radio and newspaper personnel, a line of showers, one that stretched from Canada to the Gulf, was going to move through town around five, but there were no severe storms in that line. To reassure people, WACO, the city's oldest radio station, called the National Weather Service. That station got the latest updates and broadcast an "All Clear" report at 4:30. They were unaware that at that very moment they were reassuring the community, a tornado had set down just outside of the Waco city limits and was tearing up fields and barns with winds of more than a hundred miles an hour. It was moving toward the heart of town at just over a half mile per minute.

At that same time, just as the evening edition of the newspaper was being published, a huge black cloud settled over the sun and bathed the city in an eerie darkness. For a moment, the storm was silent. Then, with a blast of lightning and a loud crack of thunder, huge drops of rain fell from the sky in sheets. Mixed in with the rain was hail, some of these icy stones as large as baseballs. Downtown people were suddenly running for cover. Hurriedly parking cars, hundreds raced into buildings to wait things out. In the Joy and Gem theaters, the rain and hail were falling so hard and fast that projectionists had to turn up the sound in order to keep the pictures from being drowned out. It was, as Texans were fond of saying, a real gully washer.

Wayne Moals felt luckier than most. He had just gotten to his car before the downpour hit. He was already five blocks away from the business district when the rain began to fall. Within seconds of seeing the first drops, he had been forced to switch on his lights and flip the knob for his wipers. Yet, even turned up all the way, his vacuum wipers could not keep the windshield clear. As Moals approached Eleventh Street all hell broke loose. His sedan was hit by a wind that literally blew his car backwards. Pushing on the accelerator did no good; even with the pedal on the floor the car and driver were pushed straight back. The wild reverse ride would last almost a block, then abruptly end. Scared to death, Moals was relieved that no one had been behind him and that little damage had been done to his car. I am a lucky man, he thought. He would not find out just how fortunate he was for several more hours.

As Moals was being pushed around by winds that now exceeded a hundred miles an hour, just a few blocks to the south and west men were putting a new roof on a house. One of the workers pointed to the black cloud that was headed toward downtown and said,

"Looks like the business district is going to get pounded." Yet because there was no visible funnel, the roofer could not have known just how understated his observation really was.

A few blocks from downtown were the tracks for the MKT railroad. Many who had ducked inside buildings or pulled over and were waiting out the rain in parked cars thought the loud roaring they now heard was coming from a freight train—maybe one of the new diesels that were replacing the old steam locomotives. Yet there was no train rolling down the tracks, only a powerful killer wind with the combined force of dozens of trains.

With the horizon hidden by three- and four-story buildings and with the sky so dark, it is little wonder that no one saw the mile-wide funnel that had set its sights on Franklin Street, Washington Avenue and Austin Avenue. After all, not even Wayne Moals had seen a funnel when it had played with his car. Even if those downtown had seen it or been warned that it was coming, by this time there was nothing anyone could have done. Each individual's fate was now in the hands of the storm.

A few blocks from the business district sat Bells Hills, a neighborhood of single-family homes, close-knit families and a school. The tornado was in a playful mood as it roared through this area. It simply twisted homes like pretzels, blew off roofs and turned over cars. Of course, it was also much weaker than it would be in two minutes.

The twister next visited the olympic-sized Sun Pool. This incredible swimming facility, one of the city's jewels, had been built on the grounds of the Cotton Palace. The old palace was gone, now only a memory of bygone days; just one wall was left to remind people of a time when cotton was the lifeblood of the area. Yet for those who loved history, cotton and the old palace grounds still meant a great deal. This was the crop that had made scores of Wacoans rich and brought an Old South–style society of debs and debutante balls to life.

On May 11; the now two-hundred-mile-per-hour winds demolished the last remaining Cotton Palace wall. It also buried Stan

Styles. He was one of six young men who had been working on the pool, readying it for the summer opening when the swirling winds hit. In an instant, as they huddled in a dressing room, the high schoolers were covered by mud, bricks and wood. Stan died, the storm's initial victim. Soon thousands of nonbelievers would come face-to-face with the "impossible" twister.

Those working at Double Manufacturing had been ready to clock out when the storm knocked the power out. They were waiting for the electricity to come back on when the tornado hit the plant. They were lucky; though damaged, the building partly survived. If the clock had been working and the employees had made it to their cars, many would have died. In this case, the storm's actions actually saved lives.

Margaret Russell was visiting Waco. Her husband had gone on an archeological dig, but because he had decided that it would be too dangerous an expedition to take his wife on, he sent her on a holiday to Central Texas. On Monday afternoon Russell had taken her nephew downtown. She parked beside the Shear Coffee Company intending to pick up some freshly ground coffee. Before Russell got out of the car, she looked up and saw bricks and roofing materials flying off the building. Ordering the five-year-old boy to the floorboard of the car, she fell and covered him. Just as she lay atop her nephew, the building caved in on top of her car. Pinned in the rubble, she asked David if he could move and find a way out of the car. By the time he managed to push the door open and get out, the storm had passed. In a few minutes David would return with some men and they would pull Russell from the wreck.

Katy Park was next in the path. This old baseball stadium was home to the Texas League's Waco Pirates. Minor league baseball was an area fan favorite and almost everyone knew the men who played for the Pirates. Up until that moment, the most important event that the old venue had ever seen was when Babe Ruth and the Yankees played an exhibition contest on the grounds more than two decades before. On this afternoon, manager Buster Chattam was not thinking about the Babe; rather, he was looking with disgust out his office window at a Katy locomotive that had sat outside his ballpark for weeks. Chattam wanted it moved and he had complained countless times to the railroad offices. He thought it was an eyesore. As he studied the big black beast now, something caught his eye. The manager watched the skies grow suddenly angry and heard a loud roar. He didn't have to glance at his outfield scoreboard, he already knew the score. Racing as fast as he could away from the park, he yelled at a groundskeeper to follow him. They literally felt the ground move as they dove and hid under the unwanted locomotive. Chattam would never again complain about the massive iron hulk. While the twister leveled Katy Park, crushing it flat and reducing the grandstands to splinters, it could not lift the locomotive. The old train saved the two men's lives.

Next, Waco's First Methodist Church lost a steeple and roof. Then a publisher was hit. In the latter, a writer for the *Waco News-Tribune* was injured, becoming the first member of the press to come directly in contact with the killer storm. In the next few seconds the Dr Pepper bottling company, one of the city's most famous landmarks, had its south wall taken apart brick by brick. At the time, Dr Pepper's slogan was 10-2-4—the tornado was a bit late for the four-o'clock drink break.

The last major building attacked before the tornado hit downtown was the Cotton Belt passenger station. It was completely dismantled. The twister then jumped the tracks and headed into the heart of the business district. Though it had been extremely powerful and had already taken a life and caused incredible destruction, its worst was just ahead. In the next three minutes this storm would end more than one hundred lives and bring a large commercial district to its knees.

Ed Daniel, who was working at a blueprint shop in the 700 block of Washington, had no idea that he was about to be hit by a powerful storm. As he worked at his desk he was oblivious to the fact that Waco was being blown up all around him. He just knew that it was raining hard and the wind had picked up. Daniel only quit work when he heard something strike hard against the side of his building. Looking out the front door, he saw a large sign that read "26th and Franklin." The sign should have been nineteen blocks from where he stood. Daniel contemplated how in the world anything that large could have blown that far. What a wind, he thought. As it turned out, this metal sign was really one of the smaller and least harmful projectiles hurled by the fierce storm.

Earl Boren, a sixteen-year-old boy, had come downtown just after lunch with his father. While his father had taken care of some business, Earl had taken in a movie. Beating his dad back to the car, Boren was sitting in the passenger seat reading the story of the tornado in Minnesota when he felt the wind shake the family car. Before he could even react, he heard the grinding of metal on metal as another vehicle was blown across the street with such force that it crushed the Boren's auto, forcing Earl to the floorboard. Trapped, unable to even turn over, Earl knew that no one could see him. He couldn't even yell loud enough to alert those close by that he was in danger. Panic and fear joined the boy on the floorboard as the car's roof groaned and collapsed another few inches toward him. Then, when it appeared as if he were going to be slowly crushed, a huge boulder slammed into the door beside his head. This large hunk of concrete, sounding like a bomb when it hit, came within inches of smashing the boy before the car did. Yet as it struck the car and bounced off, the concrete tore a hole in the metal that miraculously provided an escape path for Boren. Covering his head with his coat, he dodged debris and crawled over the rubble to a nearby building.

Just down the street, Jerry Jordan, a Baylor University student, was justly proud of his new Studebaker. He was showing it off to some friends when he noted the rain and felt the first few hail stones. He hurriedly pulled the car up against a building for protection. As if angered by the youth's actions, the tornado ripped the warehouse to pieces, pelting the new car with bricks. A whole wall blew over and crushed the car just as Jordan opened the car door and escaped into the pouring rain.

As Jordan ran for his life, Simon Guerra, who had spent the day on jury duty, had just parked at Sixth and Franklin. He didn't like the way the buildings were swaying in the wind and decided to move. Starting his sedan, he pulled away from the curb just as a building began to be pulled apart. As he hit the gas and sped away the storm began to toss bricks at Guerra. Ultimately, all the cars but his were buried. If he had delayed just one more second, he would have died.

At first those in buildings seemed more fortunate than those in cars. As cars became airborne, those watching through windows were happy they were safe inside. But this tornado was not just another twister that could only tear apart weak frame structures and toss around unsecured objects. It had power that no one in this town, and perhaps no one in the world, had ever seen firsthand and lived to tell about.

At first buildings only twisted, their windows blowing out or their roofs being lifted from their frames. Then the tornado began to pull bricks away from the top floors. Finally, as the center surged into downtown, the real force began to exert itself. Still, even as buildings began to break apart, no one realized that this was a tornado, much less the most powerful tornado in the history of Texas.

At the Joy Theater the electricity began to go on and off. With the projector starting and stopping, some of the patrons began to get concerned about what was happening outside. Just before the storm hit the movie house, the lights went off and stayed off. Now completely in the dark, one hundred people heard an unholy roar. An airman from Connally Air Force Base yelled, "Hit the deck." A second later, the roof disappeared and for an instant people could see the dark sky and feel the pounding rain. Then the storm released the roof and pushed it back onto the theater floor. A screaming mass of humanity crawled toward the lobby as they were struck by the very thing that had been holding back the rain. Still, compared to those next door, they were very lucky.

Across the street at the Dennis Building, a man who had somehow not noticed the rain that had now been falling for an hour suddenly realized that he had left his car windows down. Racing from the store in a downpour, he managed to jump in the car at just the same moment the storm brought its powerful hand against the furniture store. His forgetfulness probably saved his life.

At the Dennis Building the wind first knocked a huge water-storage tank over and sent it rolling off the roof. When those in the building heard the ominous sound of the rolling tank, they froze in their tracks. With the power off and skies dark, the only illumination was from lightning. Suddenly windows were blown out in rapid-fire progression. Beatrice Ramierz, an employee, only a year out of high school, worked on the second floor. Around her employees and customers were running in every direction. In truth they had no place to run, no place to hide, no way to reach safety. Within ten seconds of pushing the water tank from its place atop the building, the tornado did the unthinkable: In one moment its powerful winds caused the walls and then the five floors of the furniture store to spiral downward. The building just caved in and Beatrice, along with scores of others, slid into a pit and were covered with rubble. Ramierz heard cries for help all around and crawled through a hole toward the cries. Like a rat in a maze she continued to crawl through the rubble looking for more holes until she found an elevator shaft. There, she worked her way carefully down, and after finding a small opening at the bottom, she made it out into the pouring rain. As she stood in the middle of a street piled high with bricks and every kind of product imaginable, she searched for those who had been in the store with her. As she frantically looked around, she saw no one. Turning back to the pile of bricks and mortar that had been one of the city's finest buildings, she considered how many were still inside.

As Ramierz thought about who might be dead, Lillie Matkin was wondering if she was really alive. Under a mattress, trapped by tons of debris, she could only move one arm. She could also see nothing and hear only the faint sound of rain falling. A switchboard operator who had worked in the building for three decades, Lillie still didn't really know what had happened. All she knew for sure was that she had a little air and very little hope.

The storm was not satisfied bringing the Dennis Building down;

it was geared up to do more damage. It pounded the ten-story Professional Building next, knocking out windows and tearing off the roof.

Earlier in the afternoon, Bobbye Bishop had taken her mother to a doctor's appointment in the Professional Building. The two women had waited in the lobby for almost half an hour during a downpour. Right before the tornado hit, the heavy rain had slowed down. Bobbye took advantage of this opportunity and made a dash out to her car. Just as she got there the wind picked up and the rain again pelted down. Though she hadn't even put her key in the ignition, she felt the car lunge forward. Instinctively she pushed the brake down. Looking up, she watched as cars, signs and bricks flew past her. Frozen with fear, she then watched in her rearview mirror as buildings crashed down on other parked cars. Suddenly, Bishop's car pitched back and forth, raised up and down, almost stood on its front bumper. Crashing back to the ground, the sedan sat there for a second. Then, as Bishop stared out the back window, another car blew through the air at a speed far above the limit for Austin Avenue. The airborne two-ton bullet smashed into Bishop's vehicle, pinning the woman and the car to the ground. Frightened out of her mind, she had no idea that the Detroit-made iron projectile that had almost killed her had probably saved her life. The storm could now not lift her car off the ground, and the other car was shielding Bobbye from the tons of debris that were filling the air.

Willie McLain was playing dominoes at a pool hall on Bridge Street. When the storm hit, he never left his chair. He died at the table, buried by the building that served as his regular haunt. In his pocket was a life insurance policy he had taken out just that day.

In the streets thirteen-hundred-pound electrical transformers flew through the air like gigantic wrecking balls. They went through brick walls like cannon shells. They flattened trucks and cars. Never in all their years of service had they conducted this much power. In the midst of this rain of terror William Garner, a middle-aged man, ran for more than a block carrying a seventy-five-pound extermination tank on his back. He never thought to drop it and lighten his load. Somehow he outran the storm, dodged tons of debris and was completely uninjured.

All around Garner show-window displays were being blown for blocks. Huge water tanks were toppled from roofs and crushed people on the street. A surprised truck driver was even lifted five stories into the air, above almost all of the buildings. From his seat he watched as pieces of downtown Waco flew in every direction around him. The tornado carried his four-ton vehicle for a few blocks, then set it down, undamaged. The truck and driver thus escaped being another of the more than two thousand vehicles destroyed in the downtown area in just three minutes. Inside a dozen of those vehicles were men and women who would never breathe again.

The outside clock at the First National Bank was a little fast. Its lighted hands pointed to 4:40. Yet those inside didn't notice the time or the fact that the power had suddenly gone off. What they saw from the bank lobby and offices were cars bouncing back and forth like Ping-Pong balls. They felt the storm's force too as the wind blew so hard that the building was rocked. Just down the street at the Alico Building, those on the top floors were not just seeing what was happening, they were getting the ride of their lives.

The old skyscraper was feeling the full force of the storm's now estimated five-hundred-mile-an-hour winds. The building was swaying so much that windows were popping out like heated popcorn, glass was hurtling through the air like bullets, and desks and chairs on the top floors were sliding from one wall to another. People were screaming, praying and begging. Scores raced down dark stairwells as they moved with the winds. Others were stuck on elevators, listening to sounds of a force like they had never heard or believed could exist in their town. And a few, those overcome with awe, were standing in broken windows, holding on to wooden frames as they watched, mesmerized, as the twister blew by them and attacked another section of town.

As the storm moved on, a parked bus was suddenly tilted then blown over on its side. Huge pecan and oak trees were pulled up by their roots and tossed through the air like toothpicks. The suspension bridge, a historic landmark built in 1870 as part of the Chisolm Trail, and one of the first bridges of that type ever constructed, swayed like a kite. Those trapped in cars on the bridge were caught in an amusement ride that was far more terrifying than any a human had ever devised. It is a testament to its design that the bridge survived at all.

If the three minutes of fury could be compressed into just one story, it would center on Gloria Dobrovolney. When she had gotten home from school, the high school student's father, a civilian worker at Connally Air Force Base, had asked if she wanted to drive him downtown to purchase a bag of bird seed for the family parakeet. Gloria agreed, as long as her best friend, Barbara Johnson, could go along too. When they arrived downtown it was just past 4:30 and the rain was falling so hard that Mr. Dobrovolney ordered Gloria to drive around the block while he dashed through the deluge into the Texas Seed Company. The weather grew even worse as the two girls circled the block. By the time they returned to the store, the wind had picked up, bricks were flying off nearby buildings and plate-glass windows were exploding. Worried about the girls, Dobrovolney waved to them from the door. "Get in here," was what he seemed to be saying, though neither of the teenagers could hear him.

Gloria told her friend to make a run for it. As Barbara raced through the rain to the door, Gloria pulled into a parking spot. She was about to get out of the car when she heard a strange rumble. Glancing up, she saw her father and friend in the entryway of the feed store waving to her. Then, in an instant, they disappeared under tons of bricks. The seed store had been destroyed in less time than it took to open a car door.

Ignoring the rain, wind and hail, Gloria forced the car door open and ran up to where she had last seen her father and Barbara. With her bare hands she tore into the fallen building. Ten minutes later, a soaked and teary-eyed Gloria was joined by others. They would find out a day later that they would have been too late even if they had been able to move the whole building as quickly as the storm had brought it down. Gloria's father and friend had been instantly crushed.

As if on a mission to make up for all the years the city had been protected by some supernatural force, the tornado made the most of its brief stay on Texas soil. It played no favorites in its brief visit to Waco either. It hit black and white, rich and poor, visitor and resident, with equal vengeance. Still, after those few minutes, it was over, gone, the skies lightened and Waco again was a tornado-free zone. As people ventured out into the streets, the only sound was a lone car horn, the car and its driver crushed by bricks from the Dennis Building. The horn would continue to blare for hours, reminding everyone that the unthinkable had happened in Waco on May 11.

As thousands surveyed the damage, just seven blocks away the Waco fire department was waiting for their next call. They had not lost a shingle, not seen a hail stone or even experienced much wind or hard rain. The men at the station had not heard or seen anything that indicated to them that an emergency needed to be addressed. When they did receive a call for aid, they were so incredulous that they sent a scout downtown rather than a whole squad. What that man saw in the thirty-block area that had been the heart of Waco left him speechless. It looked like London during the height of the bombing runs of World War II. It was as if nothing was left. Bloody victims were everywhere and water was running through streets in torrents. Live power lines were lying on sidewalks, gas mains were broken and city water lines had ruptured and were blowing water fifty feet into the air. Shocked people were milling around, crying, begging for help, babbling, yelling and looking as though they had just walked through hell. My God, he thought, we cannot even begin to handle this. Getting on the phone, he put the word out to alert all area hospitals and call out the troops from local military bases.

One of those who walked by the inspecting fireman that day was Susie Lane. The seventy-year-old woman had been in the Joy Theater when it collapsed. Though uninjured, it took the lady more than an hour to find a path through the rubble and make the five-block walk home. It was a walk toward nothing; her home was simply not there anymore. The storm had wiped out a record of seventy years of life, every photo, every book, every piece of furniture, every item Susie cherished had been taken from her.

Like Miss Lane, Johnny Boyett, an eighteen-year-old high school student, had been in the Joy, but when he crawled out of the theater, he didn't leave. For the next forty-eight hours the boy helped dig out victims. He didn't quit until he collapsed and was taken to a hospital.

There were scores of buried victims waiting for Johnny and the hundreds of others who came downtown to help. Some, like Ted Lucenay, were lucky. He was taken from the fallen Padgitt Building at 8:00 P.M., alive. Many who were not found in the first few minutes breathed their last breaths in the depths of the rubble.

Stan Williams, an all-American football star, was downtown when the storm hit. Like Johnny Boyett, Williams stayed. In the pouring rain the gridiron star not only found Ted Lucenay, but more than twenty others.

Within half an hour students from Baylor University were helping police and firemen dig through the rubble. Using shovels and bare hands, thousands worked in the pouring rain trying to answer the hundreds of faint calls for help that were coming from under tons of bricks and boards. After a few hours, loudspeakers were brought in and microphones were lowered into debris. Then, waiting in silence, those gathered listened for the thin sound of a human voice.

Rescuers tunneled to more than twenty survivors in Torrance's Recreation Hall. Many were saved by the pool tables keeping bricks and the roof from crushing them. As time dragged on, oxygen was piped in for those who could be heard far below, tapping on the tables with cues and balls. Some had limbs amputated to free them from the wreckage.

"My husband is down there," one lady said as she watched the rescue efforts. That chilling statement was repeated scores of times during the night.

At the Dennis Building things were incredibly grim. There was nothing left, not a single portion of any of the store's five-story walls was standing. Twenty-two bodies were pulled from the wreckage. When all hope was gone, a full eighteen hours after the storm had wreaked havoc, Lillie Matkin was found, alive and uninjured.

The local hospitals were filled within the first two hours. Nurses found out names and addresses, wrote them on tape and placed them on the foreheads of each new patient brought into the emergency room. For many of the staff in the hospitals who had seen action in World War II, the emergency rooms looked like a war zone staffed by MASH units. Soon Providence and Hillcrest, though both large hospitals, could not handle any more of the two thousand suffering major injuries, so they were sent to the facility at the Connally Air Force Base and to the local Veterans Hospital. Most were transported in private cars since the ambulance companies were as overwhelmed as the hospitals.

As dawn broke on May 12, the real picture of destruction was brought into sharp focus. All buildings around the City Hall Square had been destroyed, though the City Hall itself, built in the 1920s, had somehow survived with only moderate damage. Bridge Street, the oldest in Waco, was devastated. Nothing was left on this once busy street but rubble. The Farmers Market had become a tomb. Like the Joy, the Gem Theater collapsed. People wondered why only thirty-seven had died. It should have been thousands. Yet, as would become quickly apparent, the final toll would quickly climb.

For three days the listening for cries and the digging through rubble continued. Many worked without sleep. And the rain never stopped. Thanks to the efforts of those tireless workers, more than five hundred people—coeds, housewives, store managers, infants, schoolchildren, bankers and city officials—were saved from the carnage during that time. Yet another 114 victims were added to the death toll.

The final totals revealed how risky the flight to buildings had been. Twenty-two were killed in the Dennis Building and seventeen across the alley at the pool hall. Fifty-six were killed in that block alone. Thirty-seven died in buildings on Bridge Street. Yet only a dozen met their ends while waiting in cars.

On May 15, the last body was unearthed, but nothing returned to normal. Eleven more inches of rain fell downtown, flooding many of the areas that had been hardest hit. It would take weeks to bring the power back on-line. Burying the dead took almost as long, even though multiple burials took place. In a strange quirk of fate, in the four days after the tornado, not a single person died of natural causes in Waco. There were only tornado-related deaths in the obituaries in the newspapers during that time.

All told, the storm that hit the tornado-proof city destroyed 346 buildings, took 114 lives and injured more than 2,000. The downtown retail center was effectively leveled in just three minutes. With no Katy Park to call home, the Waco Pirates took their bats and balls and moved to Longview. Most of the downtown businesses moved to the outskirts of town. And the Dennis Building became a vacant lot.

Five decades later downtown Waco is a shell of what it was before May 11, 1953. Many parts of the area were never rebuilt. For this Texas city, each dark cloud still evokes memories of the day that death came roaring from the sky with a fury unlike any an American city had ever known from a single tornado. No longer does Waco claim immunity from tornadoes and no longer do its citizens feel safe on rainy days.

# 13

## The USS *Thresher* 1961

The number-one fear that plagues most Americans is the fear of drowning. Something about dark water conjures up horrible nightmares. For that reason many stay away from rivers, lakes and oceans, afraid to venture out on them even in a large ship. It seems that for all of its beauty, millions still look upon the sea with great apprehension. It has been that way for as long as history has been recorded.

Yet in spite of the hidden dangers found in the sea, there are many who love salt air, the waves and the adventure offered in the dark waters. The mysteries of the unknown beckon them to come into the water, experience its beauty and its thrills. Without these hearty souls nations would have been bound on every shore and would never have known the world beyond. But danger and the unknown have been a draw to many. Perhaps that is what brought Columbus to America or even the men of the *Thresher* to their boat.

One of the most dangerous jobs on or in the ocean is submarine duty. It takes a special kind of person to endure the claustrophobic life of diving undersea and living in an oversized, coffinlike vessel for hours or even weeks without seeing the sunlight or breathing fresh air. Living almost shoulder to shoulder with scores of shipmates, having little privacy and with no place to escape the realities of the dangerous work, a submariner has to be a little different from every other sailor. To cruise below the sea it takes not only courage but a strong sense of personal well-being—faith, if you will, in yourself, all of those who work with you and all of those who built and service your ship.

As the sinking of the Russian submarine *Kirsk* on August 12, 2000, reminded us, seemingly routine events can go terribly wrong and no amount of faith can save us from some disasters. Around the world, people were fascinated and horrified as they learned the fate of the 118 men aboard the doomed vessel that was disabled on the ocean floor after a series of explosions. In all likelihood, the news coverage evoked all those visceral fears of the terrors of the deep that many people share. Long before the *Kirsk* went down, however, the U.S. was gripped by its own similar tragedy.

On August 3,1961, Rear Admiral L. R. Daspit of the United States Navy looked at the USS *Thresher* SNN-593, the greatest, most powerful and fastest submarine ever constructed and declared, "None of the world's oceans can be denied to her, even though she goes alone." As the sleek new nuclear sub left the harbor that day, its crew felt as if they were on the safest vessel ever designed to go undersea. There were backup systems that even backed up the backup systems, and this was a boat that seemed roomy and spacious compared to the vessels on which many of these sailors had trained. If there ever was a luxury submarine, this was it.

The *Thresher* was named for a shark. Sleek and fast, the fifteen-foot fish was well known around Cape Cod. It was not a manhunter, but the thresher was still the true master of its domain. Using its speed to run down prey and its tail as a weapon, it terrorized all the smaller fish who were unfortunate enough to swim into its path. And unlike the sub that shared its name, this shark traveled in schools, so the waters around it often turned red with the blood of the kills. The thresher shark was indeed a shimmering monster, a reason many people were scared to go near the water.

Before the nuclear submarine USS *Thresher,* another sub had sailed under the name. An underwater assassin used in the Pacific during World War II, the first *Thresher* had been a gallant fighter, but it could not compare to this new one. Dwarfed in size, power and technology, the first *Thresher* was a diving Model T compared to its namesake, which was more like an Indy 500 race car.

Built at the Portsmouth Naval Shipyard in Kittery, Maine, the *Thresher* was the first of a modern class of submarines. Nothing quite like it had been built before. The boat was able to dive deeper and run longer than any other sub in the world. It was also so quiet that few ships could even monitor its movements. In every sense, it was a man-made shark and, in the middle of the Cold War, it made America's enemies very nervous. With the *Thresher* on patrol, many veteran Russian sailors suddenly developed aquaphasia.

Like other state-of-the-art ships, such as the *Titanic,* the *Thresher* had an air of invincibility about it. It was thought the sub could survive in conditions that would doom most other oceangoing vessels. Its many backups meant that a simple accident could not take her down. Even human error could be compensated for. As a result, seamen and officers lined up to volunteer to serve on the *Thresher.* Membership on this crew was an honor that brought immense bragging rights. In an era when there often weren't enough qualified men to fill much-needed security positions, the navy had a waiting list of its very best wanting to serve on the *Thresher.*

Though it was commissioned in August of 1961, the underwater boat was really launched on July 9, 1960. Following more than a year of trials to fully check out its equipment and train a crew not used to such sophisticated equipment, it finally saw duty along with a number of other vessels in a naval exercise off the northeastern coast of the U.S. in September 1961. The ship and its crew performed well. Still, it was meant to sail alone, pushing through the dark waters as a lobo, a singular hunter looking for the most vicious prey. So the crew wanted a chance to take their boat into the open sea without escorts. This could only happen after one more round of tests.

In October, the *Thresher* left America's northeast coast for the first time and headed out to sea. It stopped at San Juan, Puerto Rico, then moved out to sea and test-fired its torpedo system. Everything was judged to be in working order. Still, after only a month at sea, the sub was ordered back to Portsmouth in November to undergo testing by designers and navy brass. America was planning on building more than twenty ships of this design, and the military wanted to get the bugs out now.

The *Thresher* would remain in port until March of the following year. It would then get to participate in another naval exercise, this time in the Atlantic off Charleston, South Carolina. After those tests, the sub docked in Florida, where it was damaged by a tug boat. The wounded ballast tank would be easily fixed and the sub would next spend some time off the Florida coast doing further tests on new equipment. Then, even though everything was working perfectly, the *Thresher* was again pulled out of duty and sent back to Portsmouth. This time it would stay in the dockyards for nine months.

In truth the modern marvel and the navy brass were at odds, frustrating its crew. It seemed that for every taste of the Atlantic, they received weeks of duty on land. This was not why they had signed up to serve on the *Thresher.* Some even decided to move onto other posts where they would taste more of the ocean. Scores asked for transfers.

For those who stayed or transferred to the ship, the winter of 1962–63 was not a time any of these proud men enjoyed. Rather than going where no one had ever gone before and causing chills to run down the enemy's back, the men were fighting snow and cold wind. If the ship was the best-designed submarine in the fleet, they grumbled, why all the down time?

In January, the *Thresher* got a new skipper. Lieutenant Commander John W. Harvey was a former football star, an Annapolis graduate and an experienced submariner. He had graduated eighth in a class of almost seven hundred. He was bright, one of the best the navy had to offer. He was also married and the father of two small children. Calm, efficient and all business, he was nevertheless a favorite of the men. Harvey knew from experience that the man who captained a submarine had to be both leader and friend; the quarters were simply too close and the duties too demanding for a leader not to be able to meet his men on all levels. Harvey could easily accomplish that. With every one of his measured steps, he inspired both confidence and trust.

For a decade, submarining had been Harvey's life. He had spent three years of that time on the most famous sub in the world, the *Nautilus*. He was part of the first crew to reach the north pole underwater. The Bronx native had also been the executive officer of the *Sea Dragon* when she made her rendezvous with the *Skate* under the polar ice cap. So he was used to life not only on the sea, but in it. During his ten years of sub duty, he had been in and seen difficult situations and he knew the limits of both his craft and his men. Though just thirty-five, he seemed the perfect choice to captain the navy's latest piece of cutting-edge technology.

Even as Harvey inspected the work the team of civilians and navy engineers performed upgrading his new ship, his own father was shaking his head. Manning Harvey, a salesman, had urged his son to abandon his naval dreams in a boat that was designed to sink. The elder Harvey didn't trust subs, and this mistrust was not comforted by his wife's saying, "Don't worry, the Lord will take care of him." It wasn't that Manning doubted God, he just didn't think God would have wanted any man to take a ship undersea. It just wasn't natural. Yet his father's words had no impact on the son. The younger Harvey felt more alive under the water than he ever did on land.

As the skipper waited for the *Thresher* to get a clean bill of health, he spent time getting to know his crew. He liked them. They were young, focused and proud to be submariners. Many of them were married, and some of them had children that hadn't even begun school. Some were from parts of the country—such as Arizona or Colorado—where the ocean was a mere afterthought. Others came from the Atlantic shores and had dreamed of sailing since childhood. And no matter where they called home, they all seemed to desire to be the best. It was that desire, combined with a lot of talent and hard work, that had brought them to Portsmouth and this sub.

Dick Powell, who often took abuse from his fellow sailors for having the same name as a famous actor, was an outgoing young man from Exeter, New Hampshire. His smile was quick and he was always ready for a joke. He had been married just over a year when he met Harvey.

Donald McCord had been with the crew for a while, and didn't take as much interest in developing a relationship with the new skipper as did many around him. This young man, also from Exeter, was tired of shore time and had put in for a transfer. McCord was more concerned about life on the *Calhoun,* his next sub, than he was about how Harvey would change the way things were done on the *Thresher.*

Billy Klier, an engineman, had been thrilled to latch onto a spot on the new sub. To work with a nuclear engine installed in the fastest fish in the navy was something he had spent his career dreaming about. His son, also named Billy, couldn't read yet, but he already wanted to be a navy man too. He didn't understand why he couldn't go along when his dad returned to the sea.

Clyde Davidson was also hoping the ship would be in the water soon, but he had other things far more pressing on his mind. He had asked Barbara Marshall, a sweetheart from Hobbs, New Mexico, to marry him. She had committed to being a navy wife, something many felt took even more courage and discipline than being a sailor. Barbara would be left alone for months at a time and have to be mother and father when the couple had children. In the face of this she had decided to put the couple's picture in the Hobbs newspaper announcing their engagement. The story was scheduled to run April 14, 1963.

Edgar Bobbitt was from the plains of Texas. While a child of the desert, he loved the ocean. His father had given his blessings for a naval career, but back in Midland the elder Bobbitt often wondered why anyone would ever want to leave the solid footing of dry land.

In Groton, Connecticut, Neil D. Shafter and his wife could only wonder how they had been so lucky as to have two sons who had made the list of about one hundred men assigned to the *Thresher.* Benjamin and John were both proud sailors, career military men in their thirties, and now even more proud because they were on the most modern and safest submarine ever built.

As Harvey met man after man, he discovered that he not only liked them, but felt secure with them. This was a good crew, one that knew its stuff and would work in a teamlike fashion when they hit the sea. The problem for the skipper, as well as his men, was that no one really knew when that would be.

Throughout the months of January and February the work crew pounded on the submarine. New electronics and sonar gear were installed. A three-foot hole was cut in the hull to put in service a state-of-the-art garbage disposal system. Now waste could be shot from the sub like a torpedo. Engines and pumps were overhauled and every switch checked, and if necessary, replaced. Some of the old piping and valves were checked as well. In March, under orders to get the boat back in operation, the work pace picked up, but still nothing was compromised. Welds were not only checked by inspectors but x-rayed too. Any flaw, no matter how slight, was reworked. Every facet of wear or tear on the equipment was logged. No ship had ever undergone scrutiny this intensive or tight. Finally, in April, with the final checks and rechecks being made, Harvey told his men it was time to stow their gear and head out to sea. The *Thresher* was again ready to be the pride of the fleet.

Monday night, April 8, 1963, the crew of the *Thresher* went to bed knowing that the next day they would finally get to do their real jobs again. The boat was going out in the deep Atlantic. Yet it wouldn't go out far or for very long. The men were assured that they would be back in plenty of time to get their taxes completed and in the mail before the deadline and also to take part in a party celebrating the sixty-third anniversary of the first submarine launch by the U.S. Navy. Still, being on the sub in the water for just two days was better than being stuck on shore.

In a very real sense, this assignment seemed silly to some of the men. It was as if regulations were getting in the way of common sense. The sub had been out many times before and each time had proven its worth. It had been tested time and time again while in drydock. A shakedown cruise did not really seem necessary and certainly wouldn't have been ordered if this had been a time of war. Everything had already been checked and rechecked, so why do it all over again?

In addition to the sailors who would be going for the ride, seventeen civilians would be making the trip too. These nonenlisted men were there to make sure that the repairs and modifications had been done properly. If there were any problems, it would be their job to address them and offer suggestions on how to correct them. They would also be allowed to judge the performance of the new equipment and help decide if the *Thresher* was again ready for active duty in the open sea or if the boat needed more time in port. For these men, who worked above the surface and on the ground, the night before the shakedown cruise was anything but a normal one. Some hardly slept at all as the excitement and the unknown crowded their minds and interrupted their dreams. For these civilians, April 9 would be anything but routine.

Thanks to an early-morning call and navy efficiency, the men and ship were ready at 8:00 A.M. Lieutenant Commander Harvey was excited as he went down the checklist with his officers. Finally he was getting to take his boat out to sea. Besides a few family members who had come to the dock to see the *Thresher* off, about the only other onlookers were aboard the *Skylark,* a navy rescue ship. That ship and its crew had been assigned the task of shadowing the big sub. They would track its course, help if there was any trouble and keep in constant communication with the fish when it dived. Like those on the *Thresher,* the men of the *Skylark* looked at this as little more than another routine operation. About the only thing that concerned them was the weather. Even though it was beautiful at the moment, the forecast indicated that the ocean might get a little rough on their second day out.

As the ship and sub readied to pull out of Portsmouth, the sixteen officers and ninety-six enlisted men went to work, and the seventeen civilians tried to observe and stay out of the way. The outsiders knew that most of the seasoned sailors didn't care for the technicians being on board. The navy men didn't like having their every move watched and recorded and they really didn't want nonmilitary types suggesting how they could best do their jobs. In their minds when the ship was at sea, it was time for navy training and sailors to take over. There was no place for anyone else on this two-day mission, or any other mission, for that matter. So the outsiders not only felt, but knew they were not really wanted. Still, even for those with a fear of being closed up and under the waves, it was exciting to watch the machine they had worked on be put to the test.

There was a formality that went along with the *Thresher*'s departure from port. As the men took their positions, a quartermaster blew his whistle. As the turbine engines began to hum louder, a flag went down astern and another went up aft. For those vets who watched this routine action on this sunny morning, chills crawled up their spines. It was always that way for ex-sub men. There was nothing like going out to sea.

As the big boat plowed through the harbor, observers had to wonder why it had been named after a shark. It looked much more like a killer whale, a giant orca, with a humped back and a sweeping tail. With no flat decks, not even an edge or corner on the hull, it was like nothing that had ever sailed the sea. It was so rounded that the crew had to wear special suction-type shoes to keep from sliding off into the water. The unique hull design allowed the *Thresher* to knife through the ocean at speeds inconceivable for those used to even other modern nuclear subs. As one of the wonders of the modern world, its beauty and power were on display as it headed out to open waters.

As they headed away from shore, the *Thresher* and her crew traveled almost over the spot where in 1939 the sub *Squalus* had gone down on its maiden voyage. To those on board the new ship, that sinking seemed as if it took place in the dark ages. Yet to a man they must have wondered what the fifty-nine men had thought as the *Squalus* suddenly sank to the bottom some four hundred feet below the surface. The crew of the *Thresher* were grateful that they lived in a different age when submarining was no longer an experimental facet of navy work, when boats were large, well tested and safe.

Four hours out to sea, the *Thresher,* with the *Skylark* looking on, sounded the klaxon and began its first dive. Burying its nose in the cold waters of the North Atlantic, the sub tasted deep salt again. For the boat and the crew, it was a good taste. For the next several hours Harvey and his men put the *Thresher* through a number of routine dives. Each new operation brought about perfect results. In spite of having a new skipper and having been on shore for nine months, the crew and the ship seemed to be functioning as one. With no problems reported, Harvey had his communications specialist alert the *Skylark* that it was time to move farther out to sea. The night would be spent getting to a point just past the Continental Shelf, where the sub could be tested in truly deep water.

Even among the crew, few knew just how far down the *Thresher* could go. That information was classified. But this sub, along with its recently launched sister subs *Permit* and *Plunger,* was said to be rigged to survive below a thousand feet. Some bragged that they could probably work at even fifteen hundred. This trio of mechanical sharks was simply by far the best in the world. They had to be. Their job was basic—seek out and kill any enemy subs in their paths. To kill them, they had to be faster, stronger and better than the enemy was. These three forty-five-million-dollar tear-drop-shaped boats were all that and more.

Tuesday had been a beautiful day, but Wednesday was a mess. The winds were whipping the seas into a fury. Nine-foot waves rocked the *Skylark* and its crew. Those on the *Thresher* smiled as they watched their escort bob in the water. They knew that once the sub dove under the angry seas, the crew would feel none of the pitching. Theirs would be smooth sailing.

On April 10, Harvey and the *Thresher* crew had one job, to test their boat in deep water. Operating over the Continental Shelf, they had plenty of room to make the tests. The clay-and-rock ocean bottom was more than eight thousand feet below the surface. That day, as the men heard the dive klaxon, they knew there was no danger of sticking the boat on the bottom. Even this super-fish couldn't go down that far.

Once again everything appeared to be normal. Staying in contact with the *Skylark* via the undersea phone, each new communiqué indicated the sub was performing perfectly. Just twenty-four hours after leaving Portsmouth, communications reported that half the tests were completed and the sub was going deeper.

On the *Skylark,* as the winds continued to buffet the ship, men manning the radio equipment and sonar were looking forward to the *Thresher* completing the shakedown. Bobbing like a cork in a tornado was not a pleasant way to spend a day. They wanted to go home.

At 7:54 A.M., the *Thresher* reported that the sub had passed four hundred feet and was leveling off to check for any leaks. They also reported that there would be no more mentions of how deep the sub would ultimately be diving. The secretive facet of the mission had now begun. It was obvious that the navy didn't want any of its Cold War enemies to figure out the diving parameters of the killer ship.

Fifteen minutes later the sub again contacted the accompanying ship and reported things were fine. Then a voice added, "proceeding at one-half test depth." The *Skylark* didn't know how deep that was, but as nothing seemed to be amiss, the rescue boat simply asked the sub to make contact again in a quarter hour.

In the next half hour, three reports came in that indicated the *Thresher* was doing fine. Everything was going as planned. On the *Skylark,* the crew stayed alert, but they remained relaxed.

In the sub, even at a greater depth, the routine of ship's operations looked little different than it had the day before. For the civilians, the popping and pinging caused by the ocean's pressure pushing against the boat must have been unnerving. It sounded as if the boat were going to be crushed by the weight of the water. Yet these machinegun-like groans were really the norm and didn't faze the crew in the least. It happened every time the ship challenged the power and strength of the sea. With all 130 of the structural modifications completed at Portsmouth holding, it appeared as if the sub was again ready to finally put out to sea on a long mission. Only one more test remained.

At 9:13, something went wrong. When the sub checked in with the *Skylark,* communications reported that there were "minor difficulties" and that the boat was going to blow out its tanks and return to the surface. Those manning the listening devices on the ship could hear the sound of an air-pressure release coming from deep under the sea. Then, they heard nothing.

Concerned that something major might be wrong, Commander Hecker of the *Skylark* took over the sea phone. He wanted to determine if there was anything the escort ship could do to help the sub with its current problems. Speaking into the phone he asked, "Are you under control?" When no one answered, he asked again, "Are you under control?"

Just a few minutes before Hecker had taken charge of the phone, those in control of the *Thresher* had confronted a problem. A crewman had noted a leak in the engine room's water piping system. The leak was a surprise, but hardly a shock. Water pressure leaks were common on most subs. Grabbing a wrench, the man set about slowing and stopping the leak. Yet rather than dissipate, the water began to spray from the connection in a steady stream. Seeing that he could not fix the problem by himself, he notified command of the leaks and asked for help.

As other crewmen joined the first man, seawater began to shoot into the sub from the piping. Not panicking, the crew did what they were trained to do, fix the leaks. Yet every time they attacked a spot, another broke out. Studying the problem, sensing that it couldn't be fixed with the boat this deep in the water and under extreme pressure, the engineers decided to address the leaks on the surface.

Communications was given the order to alert the *Skylark* of the situation and that the *Thresher* would be coming up.

As Commander Hecker listened, a garbled voice came over the line. It was 9:17. The skipper and others listening in could not make out any of the message except the final two words, "test depth." What does that mean? Hecker wondered. Looking around at his communications crew, he was met by faces that reflected his own confusion.

While they continued to try to contact the sub with the phone, the skipper notified the bridge to scan the ocean's surface for the *Thresher*. Logic told him that communications had simply gone out and that the sub would be surfacing momentarily. Besides, orders had been to only contact the *Skylark* every fifteen minutes, so there was no real reason to worry if they didn't immediately respond.

Back under the sea, the fight to cut the leaks was losing ground. More were springing up. The piping joints, made of a silicon-brass compound, were giving way. As they did, something inconceivable happened. The flow of water shorted out the *Thresher*'s electrical system and somehow shorted out the nearby backup system as well. Then as part of the safety procedures, the nuclear reactor shut down. Within seconds the ship was dead in the water. Without power, the sub hovered a thousand feet beneath the ocean surface and everything was deathly quiet. Then, realizing the peril they were in, frantic men attempted to find a solution and get the power back on-line.

The *Thresher*'s crew had to find a way to blow the tanks and get to the surface. Without power, there was no way. Trying to restart the failed electrical system that had plunged the ship into darkness was the initial focus of the officers, the mechanics, engineers and civilian technicians. While they labored, the rest of the men waited. They listened to scores of suggestions and experts scramble to different positions. The ship groaned and popped as the power of the Atlantic began to push more heavily against its hull. Now, even to the most experienced crewman, those noises meant something.

Too deep to use the electrical propulsion motor and unable to refire the nuclear power grid, the *Thresher* was in a life-or-death struggle to somehow fight its way back to the surface. The complete loss of power meant that the crew couldn't empty the ballast tanks because they were electric. Now they had no other choice but to manually blow the tanks and hold on for the rough ride topside. The order to blow was given. This was the air pressure sound that those listening electronically on the *Skylark* had heard. *Now,* the men in the sub thought, *now we can get out of this mess. The final backup to the backup has saved us.*

As the seconds ticked by, it became obvious that something was dramatically wrong. Rather than rise, the ship began to sink deeper. The tanks hadn't blown their water. Every one of the 129 on the sub knew that if the ballast tanks were not emptied in the next few minutes, the *Thresher*'s hull would be crushed. Frantically the crewmen tried and retried to blow the water from the tanks, but nothing they did worked. Suggestions were made, new measures were tried, but the seawater remained in the tanks. Though everyone on board knew that this was impossible, that this failsafe never failed, the water remained, locked tightly inside the boat's hull.

What the men didn't know was that the manufacturer of the emergency blowing system had put strainers on the blow valve to protect them from sand particles. The process of blowing caused moisture to form on the strainers. Because of the rapid decrease in pressure and the depth of the ocean, the moisture turned to ice, forming a seal over the drain valves. This ice clogged the valves and prevented the blow from emptying the ballast tanks. Only a little air escaped. So the ship continued to sink as men prayed, cried and begged for the valves to let the water blow.

Within twenty minutes of the crewman discovering the initial leak, the *Thresher*'s hull began to groan more loudly. Sitting in darkness, breathing foul air, all aboard knew what would happen next. It was just a matter of time. Within maybe another minute the metal gave way to the sea's pressure and the boat and crewmen were crushed as the Atlantic rushed in like a tidal wave and broke the

*Thresher* into six large hunks of twisted metal and thousands of other pieces of big debris.

On the surface, the *Skylark* continued to use every means available, including Morse code, to try to contact the submarine. The bridge crew of the escort ship swept the rough ocean looking for signs of the *Thresher* coming to the surface. They heard and saw nothing. One seaman, manning an underwater listening post, did report hearing a dull thud. It was the *Thresher* breaking up.

At 11:04, Hecker notified the submarine base at New London, Connecticut, that the sub had been out of contact for an hour and forty-seven minutes. The navy immediately began to scramble planes to begin a search of the area. When the communication gap hit six hours, a full alert was sounded and President John Kennedy was notified.

By nightfall five destroyers, two other submarines, a submarine rescue ship and numerous other boats were in the area. Back in Portsmouth, a navy chaplain was called in and the families of the men on the sub were being notified that the *Thresher* was missing.

Though many in the navy had given up hope when they learned that the *Skylark* had heard a dull thud, a few held on to the belief that nothing too serious could have gone wrong with a sub that had so many backup systems. They figured that it was either stuck under the water, with communications down, and making repairs, or that it had surfaced and had yet to be spotted. Yet as time dragged by and midnight approached, those hopes dimmed considerably.

Back at the base, as families gathered to speak with navy officials, some remembered the fatal first voyage of the *Squalus*. It had gone down and was thought crushed at the bottom of the ocean. Yet the ship was found and more than half the crewmen were saved using a mechanical diving bell. The boat was even brought back to the surface and rebuilt. It served throughout World War II. Maybe the *Thresher* was simply stuck on the bottom waiting to be found and rescued.

Navy personnel downplayed that possibility. The *Squalus* had gone down in four hundred feet of water. The *Thresher* would not have hit bottom until eighty-five hundred feet. Rescues could not be made at that depth, not withstanding the fact that the pressure of the ocean at more than a mile and a half would crack the boat apart like an eggshell.

In the depths of the night, the destroyer *Washington* reported finding red and yellow gloves, bits of plastic and an oil slick. The gloves were from the men on the *Thresher.* At that point everyone knew that the impossible had happened. The most advanced ship in the world had somehow sunk and those on it had died.

As the news hit Portsmouth, seven navy officers, all with submarine experience, began to call the families of those who had left port just a day and a half earlier. Meanwhile, a formal statement was issued: "Admiral Anderson has concluded with great regret and sadness that this ship with 129 fine souls aboard is lost."

At dawn a group of marines marched to the Portsmouth flag mast. With drums sounding and bugles blowing, they raised the flag to the top of the pole, and then lowered it halfway down. As they did, President Kennedy told the press, "The courage and dedication of these men of the sea, pushing ahead into the depths to advance our knowledge and capabilities is no less than that of their forefathers, who led the advance on the frontiers of our civilization."

As family members stared out at the Atlantic, as they realized that they would never again see their loved ones or even get to bury them, the president's words offered little comfort. Young widows, tiny children, fathers and mothers and brothers and sisters all questioned how it could have happened. No doubt Lieutenant Commander Harvey's father was reminded that he had told his son that submarines were simply not safe.

Those who died on the *Thresher* had often bragged about how many times each system on the sub had been tested and retested. They spoke confidently of the long line of backups that protected them from any possible calamity. Yet what none realized was that the navy had only tested 20 of the 145 joints that held the water pipes together. They should have tested and retested them all. They should have considered what air pressure blown at extreme depths would do to their newly installed sand-filter grids. For the crew of this ship, who had waited nine long months to get their sub updated, it wasn't long enough and, though none of them would have believed it until the very last moment, the tests were not thorough enough.

Dying in a dark, underwater coffin remains one of man's worst nightmares. For the men on the *Thresher,* 129 individuals who had been told they were on the safest vessel in the navy, it was a nightmare come true.

# Acknowledgments

The author would like to thank the following people and organizations:

Kathy Collins
Dale Miller
Jessika Stratton
John Hillman
Madeleine Morel
Gary Brozek
Bill Harns
The Hillsboro Library
The Baylor University Library
The Baylor University Oral History Department
The United States Library Exchange Program

# Index